Studies in Computational Intelligence

Volume 638

Series editor

Janusz Kacprzyk, Polish Academy of Sciences, Warsaw, Poland
e-mail: kacprzyk@ibspan.waw.pl

About this Series

The series "Studies in Computational Intelligence" (SCI) publishes new developments and advances in the various areas of computational intelligence—quickly and with a high quality. The intent is to cover the theory, applications, and design methods of computational intelligence, as embedded in the fields of engineering, computer science, physics and life sciences, as well as the methodologies behind them. The series contains monographs, lecture notes and edited volumes in computational intelligence spanning the areas of neural networks, connectionist systems, genetic algorithms, evolutionary computation, artificial intelligence, cellular automata, self-organizing systems, soft computing, fuzzy systems, and hybrid intelligent systems. Of particular value to both the contributors and the readership are the short publication timeframe and the worldwide distribution, which enable both wide and rapid dissemination of research output.

More information about this series at http://www.springer.com/series/7092

Naoki Fukuta · Takayuki Ito
Minjie Zhang · Katsuhide Fujita
Valentin Robu

Editors

Recent Advances in Agent-based Complex Automated Negotiation

 Springer

Editors
Naoki Fukuta
Shizuoka University
Hamamatsu, Shizuoka
Japan

Takayuki Ito
Nagoya Institute of Technology
Nagoya
Japan

Minjie Zhang
School of Computer Science
University of Wollongong
Wollongong, NSW
Australia

Katsuhide Fujita
Tokyo University of Agriculture and
    Technology
Tokyo
Japan

Valentin Robu
Institute of Sensors, Signals and Systems
Heriot-Watt University
Riccarton
UK

ISSN 1860-949X   ISSN 1860-9503   (electronic)
Studies in Computational Intelligence
ISBN 978-3-319-80775-1   ISBN 978-3-319-30307-9   (eBook)
DOI 10.1007/978-3-319-30307-9

Printed on acid-free paper

This Springer imprint is published by Springer Nature
The registered company is Springer International Publishing AG Switzerland

Preface

This book includes selected revised and extended papers from the 7th International Workshop on Agent-based Complex Automated Negotiation (ACAN2014), which was held in Paris, France, in May 2014.

Complex automated negotiations include various key issues to realize intelligent agent-mediated negotiations in realistic scenarios. Since these scenarios are complex and complicated, various techniques, models, theories, and methods that came from game theory, economics, intelligent control systems, heuristic algorithms, and interactions among humans have been applied to solve them effectively and efficiently.

Researchers are actively exploring these issues from different communities in autonomous agents and multiagent systems. They are, for instance, being studied in agent negotiation, multi-issue negotiations, auctions, mechanism design, electronic commerce, voting, secure protocols, matchmaking and brokering, argumentation, cooperation mechanisms, uncertainty modeling, distributed optimization, and decision making and support systems, as well as their application areas. The goal of our workshop is to bring together researchers from these various communities to learn about one author's approaches to the problems, form their long-term collaborations, and solve these complex problems by applying and integrating knowledge studied in these different areas to accelerate progress toward scaling up to larger and more realistic applications.

ACAN has been closely and tightly cooperating with Automated Negotiating Agents Competition (ANAC), in which automated agents that have different negotiation strategies and are implemented by different developers compete against one another in different negotiation domains. Based on the great success of ANAC2010, ANAC2011, ANAC2012, and ANAC2013, ANAC2014 was also held within the International Conference on Autonomous Agents and Multiagent

Systems (AAMAS) 2014 in Paris. This book also includes an ANAC special part, where authors of selected finalist agents explain the strategies and ideas used in them.

Hamamatsu, Japan   Naoki Fukuta
Nagoya, Japan   Takayuki Ito
Wollongong, Australia   Minjie Zhang
Tokyo, Japan   Katsuhide Fujita
Riccarton, UK   Valentin Robu

Contents

Contributors

Samir Aknine Université Claude Bernard Lyon 1, Villeurbanne, France; Université de Lyon, Lyon, France

Bedour Alrayes Department of Computer Science, Royal Holloway, University of London, Egham Hill, Egham, UK

Souhila Arib Université Paris Dauphine, Paris, France

Reyhan Aydoğan Computer Science Department, Özyeğin University, Istanbul, Turkey

Tim Baarslag Interactive Intelligence Group, Delft University of Technology, Delft, The Netherlands; Agents, Interaction and Complexity Group at the University of Southampton, Southampton, UK

Roghayeh Barmaki Department of EECS, University of Central Florida, Orlando, USA

Rahmatollah Beheshti Department of EECS, University of Central Florida, Orlando, USA

Chantal Berdier EDU Laboratory Insa of Lyon, Lyon, France

Romain Caillere Université de Lyon, Lyon, France

Siqi Chen University of Liverpool, Liverpool, UK

Dave de Jonge IIIA-CSIC, Campus de la UAB S/n, Bellaterra, Catalonia, Spain

Claudia Di Napoli Istituto di Calcolo e Reti ad Alte Prestazioni—C.N.R., Naples, Italy

Dario Di Nocera Dipartimento di Matematica, University of Naples "Federico II", Naples, Italy

Eden Shalom Erez Ariel University, Ariel, Israel

David Festen Delft University of Technology, Delft, The Netherlands

Katsuhide Fujita Faculty of Engineering, Tokyo University of Agriculture and Technology, Tokyo, Japan

Naoki Fukuta Graduate School of Informatics, Shizuoka University, Hmamamatsu, Japan

Thomas Genin Hedera Technology, Montreuil, France

Rafik Hadfi Department of Computer Science and Engineering, Graduate School of Engineering, Nagoya Institute of Technology, Nagoya, Japan

Koen Hindriks Interactive Intelligence Group, Delft University of Technology, Delft, The Netherlands

Takayuki Ito Techno-Business Administration (MTBA), Nagoya Institute of Technology, Aichi, Japan; Department of Computer Science and Engineering, Graduate School of Engineering, Nagoya Institute of Technology, Nagoya, Japan; Master of Techno-Business Administration, Nagoya Institute of Technology, Nagoya, Japan; Nagoya Institute of Technology, Nagoya, Japan

Catholijn Jonker Interactive Intelligence Group, Delft University of Technology, Delft, The Netherlands; Man Machine Interaction Group, Delft University of Technology, Delft, Netherlands

Yoshiaki Kadono Graduate School of Informatics, Shizuoka University, Hmamamatsu, Japan

Özgür Kafalı Department of Computer Science, Royal Holloway, University of London, Egham Hill, Egham, UK

Shinji Kakimoto Faculty of Engineering, Tokyo University of Agriculture and Technology, Tokyo, Japan

Yan Kong School of Computer Science and Software Engineering, University of Wollongong, Wollongong, Australia

Nasser Mozayani School of Computer Engineering, Iran University of Science and Technology, Tehran, Iran

Faria Nassiri-Mofakham Department of Information Technology Engineering, University of Isfahan, Isfahan, Iran

Makoto Niimi Department of Computer Science, Nagoya Institute of Technology, Nagoya, Japan

Fenghui Ren School of Computer Science and Software Engineering, University of Wollongong, Wollongong, NSW, Australia

Silvia Rossi Dipartimento di Ingegneria Elettrica e Tecnologie dell'Informazione, University of Naples "Federico II", Naples, Italy

Yoshihito Sano Graduate School of Informatics, Shizuoka University, Hmamamatsu, Japan

Motoki Sato Nagoya Institute of Technology, Nagoya, Japan

Carles Sierra IIIA-CSIC, Campus de la UAB S/n, Bellaterra, Catalonia, Spain

Marta M. Skarżyńska Delft University of Technology, Delft, The Netherlands

Kostas Stathis Department of Computer Science, Royal Holloway, University of London, Egham Hill, Egham, UK

Bálint Szöllősi-Nagy Delft University of Technology, Delft, The Netherlands

Karl Tuyls Department of Knowledge Engineering, Maastricht University, Maastricht, The Netherlands

Gerhard Weiss School of Computer and Information Science, Southwest University, Chongqing, China

Dayong Ye School of Computer Science and Software Engineering, University of Wollongong, Wollongong, Australia

Farhad Zafari Department of Information Technology Engineering, University of Isfahan, Isfahan, Iran; Faculty of Science, Engineering and Technology, Swinburne University of Technology, Melbourne, VIC, Australia

Jihang Zhang School of Computer Science and Software Engineering, University of Wollongong, Wollongong, NSW, Australia

Minjie Zhang School of Computer Science and Software Engineering, University of Wollongong, Wollongong, NSW, Australia

Shuang Zhou School of Computer and Information Science, Southwest University, Chongqing, China

Inon Zuckerman Department of Industrial Engineering and Management, Ariel University, Ariel, Israel

Part I
Automated Negotiation Strategy and Model

Prediction of the Opponent's Preference in Bilateral Multi-issue Negotiation Through Bayesian Learning

Jihang Zhang, Fenghui Ren and Minjie Zhang

Abstract In multi-issue negotiation, agents' preferences are extremely important factors for reaching mutual beneficial agreements. However, agents would usually keeping their preferences in secret in order to avoid be exploited by their opponents during a negotiation. Thus, preference modelling has become an important research direction in the area of agent-based negotiation. In this paper, a bilateral multi-issue negotiation approach is proposed to help both negotiation agents to maximise their utilities under a setting that the opponent agent's preference is private information. In the proposed approach, Bayesian learning is employed to analyse the opponent's historical offers and approximately predicate the opponent's preference over negotiation issues. Besides, a counter-offer proposition algorithm is integrated in our approach to help agents to generate mutual beneficial offers based on the preference learning result. Also, the experimental results indicate the good performance of the proposed approach in aspects of utility gain and negotiation efficiency.

1 Introduction

Automated negotiation is a form of decision making that agents jointly explore possible solutions to reach an agreement [1, 3, 6, 8]. One significant research area in automated negotiation is to design negotiation approaches that can help agents to achieve mutual beneficial agreements during negotiation. In order to achieve win-win agreements, agents usually need to discover or predict their opponents' private information, such as utility functions, concession strategies, negotiation deadlines,

J. Zhang (✉) · F. Ren · M. Zhang
School of Computer Science and Software Engineering,
University of Wollongong, Wollongong, NSW, Australia
e-mail: jz718@uowmail.edu.au

F. Ren
e-mail: fren@uow.edu.au

M. Zhang
e-mail: minjie@uow.edu.au

© Springer International Publishing Switzerland 2016
N. Fukuta et al. (eds.), *Recent Advances in Agent-based Complex Automated Negotiation*, Studies in Computational Intelligence 638,
DOI 10.1007/978-3-319-30307-9_1

3

reservation prices, preferences and so on [5]. Among agent's private information, preference on negotiation issues is the important information that used for issues trade-off. Through trading off issues, there is very high possibility that agents can reach win-win agreements when the negotiation ends [1, 3, 7, 8].

Precisely speaking, in multi-issue negotiation, agents usually have different preferences over different negotiation issues. A high preferred issue can help agents to generate more utility comparing with a low preferred issue. During a multi-issue negotiation, the offer that an agent proposed should not only maximise its own utility, but should also not decrease its opponent's utility, thus the opponent agent would be more willing to accept the offer. In order to propose such a suitable offer, agents need to know their opponents' preferences on negotiation issues. According to the opponent's preference, agents can trade off issues and propose mutual beneficial offers [3]. However, under normal situation, self-interest agents will not be willing to reveal their preference information. Thus, preference learning has become a key factor to contribute to the win-win negotiation results [2, 3, 8, 11].

In this paper, a bilateral multi-issue negotiation approach is proposed to improve the negotiation results. In our negotiation approach, Bayesian learning algorithm is employed to predict the opponent's preference. The major contributions of the proposed approach are that (1) the proposed preference learning algorithm does not require any extra information about the opponent to initialise the learning. The learning procedure is only based on the analysis of opponent's counter-offers; and (2) the proposed negotiation approach has integrated a counter-offer proposition algorithm, which is capable of trading issue effectively based on the preference learned from the opponent. Thus, both agents can increase their utilities from the mutual beneficial offer.

The rest of this paper is organised as follow. Section 2 presents the detail of our negotiation approach, which includes preference learning and counter-offer proposition. Section 3 shows the experimental result of our negotiation approach. Section 4 discusses some related work on multi-issue negotiation. Section 5 gives the conclusion and further work.

2 Negotiation Approach Based on Bayesian Learning

In our negotiation approach, certain assumptions are made. First, both agents should have target utilities and negotiation deadlines. Second, there is no dependency between negotiation issues. Third, both agents follow the concession-based negotiation strategy to decrease their target utilities [4]. During each negotiation round, agents first check whether current round has exceeded their negotiation deadlines. If the deadline has not been reached yet, agents will concede their target utilities based on parameters defined in their concession strategies, then use utility functions to calculate an offer's payoff. Based on the calculation result, agents can accept the offer, or reject the offer and propose a counter-offer.

The rest of this section is divided into four subsections. Section 2.1 introduces the basic negotiation model and some technical terms used in our negotiation approach. Section 2.2 presents the detail of concession strategy used by both agents. Section 2.3 describes how to use the Bayesian learning to model the opponent's preference. Section 2.4 introduces the procedure of issue trade-off and counter-offer proposition.

2.1 The Basic Negotiation Model

Our negotiation model partially employs the multi-issue negotiation model proposed by Faratin et al. [4].

Let i represent one of the negotiation agent and i' represent its opponent agent and $j (j \in 1, \ldots, n)$ is one of the issues that are negotiated between the two agents. Let $x_j = [min_j, max_j]$ be a value of issue j and min_j, max_j represent the range of x_j. Each agent has an evaluation function $E^i_j : [min_j, max_j] \rightarrow [0, 1]$ that evaluates the value of issue j to a normalized value between 0 and 1. For different negotiation issues, agent $i's$ evaluation function could be different. Let w^i_j donate the issue j's weighting of agent i.

According to the above terms, an agent's utility function can be defined by Eq. (1):

$$U^i(X) = \sum_{j=1}^{n} w^i_j E^i_j(x_j), \tag{1}$$

where x_j represents the value of issue j; E^i_j represents the evaluation function that used by agent i to evaluate issue j and w^i_j represents the issue j's weighting of agent i.

In our negotiation model, we assume that both agents use the linear additive utility function to evaluate an offer's payoff and most of the issues negotiated between agents are conflict issues. Conflict issues here mean increasing the value of an issue will help agents to raise their utilities but to decrease its opponent's utility. For example, when two agents negotiate over the price for a service, the seller agent would be happy to increase the price while the buyer agent would not delight when the price raises. Therefore, the opponent i''s evaluation function on issue j can be simply assumed as $1 - E^i_j(x_j)$ by agent i and the opponent i''s utility function can be expected by agent i as:

$$U^{i'}(X) = \sum_{j=1}^{n} w^{i'}_j (1 - E^i_j(x_j)), \tag{2}$$

where $(1 - E^i_j(x_j))$ represents agent i's expectation on opponent i''s evaluation result of issue j and $w^{i'}_j$ represents agent i's expectation on opponent i''s weighting on issue j.

Let $x^t_{i\to i'}$ represent the vector of values (offer) proposed by agent i to its opponent i' at time t and $x^t_{i\to i'}[j]$ represent the value of issue j in the offer. Let t^i_{max} represent the deadline for agent i to complete the negotiation. Agent i also has a target utility $V^i (V^i = [0, 1])$, which is used to determine whether to accept an offer.

In each round of negotiation, after agent i receives an offer from its opponent i', agent i will first compare current time t with t^i_{max}. If $t > t^i_{max}$, it means the current negotiation time has already exceeded the negotiation deadline, thus agent i will reject the offer and quit the negotiation. However, if $t \leq t^i_{max}$, agent i will first concede its target utility V^i according to the concession strategy and then evaluates the offer $x^t_{i'\to i}$ by using its utility function $U^i(x^t_{i'\to i})$. The calculation result is used to compare with the target utility V^i. Agent i will accept offer $x^t_{i'\to i}$ if its utility is greater than the target utility V^i. Otherwise agent i will reject the offer and try to learn opponent i' preference. Finally, agent i will propose a counter-offer $x^t_{i\to i'}$ to opponent i'. The detail of the negotiation procedure is described in Algorithm 1 as follows.

Algorithm 1 Negotiation Procedure

1: receive offer $x^t_{i'\to i}$
2: **if** $t > t^i_{max}$ **then**
3:     quit negotiation
4: **else**
5:     concede V^i
6:     calculate $U^i(x^t_{i'\to i}) = \sum_{j=1}^{n} w^i_j E^i_j(x^t_{i'\to i}[j])$
7:     **if** $U^i(x^t_{i'\to i}) \geq V^i$ **then**
8:         accept offer $x^t_{i'\to i}$
9:     **else**
10:         learn and update the estimation on opponent i' 's preference
11:         propose counter-offer $x^t_{i\to i'}$
12:     **end if**
13: **end if**

In Lines 2–3, agent i checks its deadline and decides whether to quit the negotiation. If the negotiation time does not exceed agent i's deadline, agent i will concede its target utility and evaluate the offer to decide whether to accept it (Lines 5–8). If agent i does not accept the offer, agent i will learn its opponent i''s preference and then propose a counter-offer (Lines 10–11).

2.2 Concession Strategy

During the negotiation, autonomous agents usually follow certain strategies to propose offers. We assume both agents use concession-based negotiation strategies [4] and the concession made by an agent is strongly related to negotiation time. With the increasing of negotiation time, both agents should consider to increase the concession degrees on the negotiation issues.

Generally, agent i's target utility is set to its maximum value at the beginning of a negotiation (usually equals to 1), when negotiation time reaches agent i's deadline, the target utility should be decreased to the minimum value that agent i can accept, which can be defined as:

$$V_t^i = \begin{cases} V_{max}^i & \text{when } t = 0 \\ V_{min}^i & \text{when } t = t_{max}^i \end{cases} \tag{3}$$

where V_t^i represents the target utility of agent i at time t and V_{max}^i and V_{min}^i represent the maximum and minimum target utility of agent i, respectively .

The agent's concession algorithm can be defined by Eq. (4):

$$V_t^i = V_{max}^i - (V_{max}^i - V_{min}^i) * (\frac{t}{t_{max}^i})^\gamma, \tag{4}$$

where the value of γ is used to change the concession strategy, which can be classified as: (1) when $0 < \gamma < 1$, agent i will make large concession at the beginning of the negotiation and make small concession when the negotiation approaches to the end; (2) when $\gamma = 1$, agent i will make a constant degree of concession through the whole negotiation; and (3) when $\gamma > 1$, agent i will make small concession at beginning of the negotiation but increase the concession degree at latter rounds of the negotiation.

In our model, we assume that agent i knows its opponent i''s deadline $t_{max}^{i'}$ and concession strategy γ.

2.3 Bayesian Learning in Negotiation

Bayesian learning is usually used to calculate the explicit probabilities for hypotheses. In the Bayesian learning system, there is a hypothesis space H which contains a set of possible hypotheses and Bayesian rule is used to determine the most probable hypothesis among them [9].

$$\text{Bayes' Rule: } P(h|D) = \frac{P(D|h)P(h)}{P(D)}, \tag{5}$$

where h is one of the hypothesis in the hypothesis space H and D is the training data; $P(h)$ is the prior probability of the hypothesis h; $P(D)$ is the probability that the training data D will be observed given no knowledge about which hypothesis h holds; $P(D|h)$ denotes the probability of observing data D given the condition that the hypothesis h holds. Finally, $P(h|D)$ is the posterior probability, which represents the probability that hypothesis h holds given the observed training data D. It reflects the confidence that hypothesis h holds after the training data D has been seen.

In the multi-issue negotiation field, the hypothesis space H can be used to represent all possible ranking of the negotiation issues of agent i, and the training data D could be the offer or the counter-offer in a particular round of negotiation [7]. After each negotiation round, agent i should update the belief of each hypothesis h in the hypothesis space H according to the latest offer.

Let H^w donate the possible ranking of the negotiation issues of opponent i' and $h_m (h_m \in \{1, \ldots, n\})$ represent one of the hypothesis (ranking of the issues) that belongs to hypothesis space H^w. The weights of negotiation issues can be normalized by Eq. (6) [7]:

$$w_j = 2 * \frac{r_j^m}{n(n+1)}, \qquad (6)$$

where r_j^m donates the ranking of issue j in hypothesis h_m and the ranking starts from the least important issue to the most important issue.

Before Bayesian learning algorithm can be applied, a uniform distribution is assigned to hypotheses in the hypothesis space H^w. During each round of negotiation, when a new offer is received from opponent i', the Bayesian rule should be used to calculate the posterior probability of each hypothesis h_m. The calculation is defined by Eq. (7):

$$P(h_m | x_{i' \to i}^t) = \frac{P(x_{i' \to i}^t | h_m) P(h_m)}{\sum_{k=1}^n P(x_{i' \to i}^t | h_k) P(h_k)}, \qquad (7)$$

where $P(h_m)$ represents the latest probability of hypothesis h_m and $P(h_m | x_{i' \to i}^t)$ represents the posterior probability of hypothesis h_m given the condition that offer $x_{i' \to i}^t$ is proposed by opponent i' and received by agent i at time t.

Term $P(x_{i' \to i}^t | h_m)$ in Eq. (7) means that when the given hypothesis h_m is hold, the probability that opponent i' proposes offer $x_{i' \to i}^t$. In order to calculate such conditional probability, the approximated utility of opponent i''s offer need to be calculated. Since our negotiation approach has assumed that the opponent i''s concession detail is public information, so it is possible to estimate opponent i''s target utility in each negotiation round. Therefore, we can apply each issue's weight ranked in each hypothesis h_m on opponent i''s utility function to estimate the utility of negotiation offer. The calculation result is used to compare with opponent i''s target utility at the current round and the hypothesis h_m, which produces an estimation utility that is most close to the opponent i''s target utility is the correct hypothesis.

Let $C_{h_m} (h_m \in \{1, \ldots, n\})$ represent number of time that hypothesis h_m is the correct hypothesis before negotiation time t, then conditional probability $P(x_{i' \to i}^t | h_m)$ can be calculated as:

$$P(x_{i' \to i}^t | h_m) = \frac{C_{h_m}}{t}, \qquad (8)$$

After calculation of each $P(h_m | x_{i' \to i}^t)$, agent i should use the calculation results to update the probability distribution of hypothesis space H^w. Finally, the hypothesis h_m, which has maximum posterior probability during this negotiation round, is applied to adjust the issue's value, and then agent i proposes a counter-offer to opponent i'. The detail of the preference learning procedure is illustrated in Algorithm 2 as follows.

Algorithm 2 Preference learning

1: apply t to concession equation $V_t^i = V_{max}^i - (V_{max}^i - V_{min}^i) * (\frac{t}{t_{max}^i})^\gamma$
2: update opponent i''s target utility $V_t^{i'}$
3: **for all** $h_m \in H^w$ **do**
4:     calculate $U^{i'}(x_{i' \to i}^t)$ using preference in h_m
5:     save result of $(U^{i'}(x_{i' \to i}^t), h_m)$
6: **end for**
7: choose h_m which generate $U^{i'}(X^t)$ that most close to $V_t^{i'}$
8: update C_{h_m} by $C_{h_m} + 1$
9: **for all** $h_m \in H^w$ **do**
10:     calculate $P(x_{i' \to i}^t | h_m) = \frac{C_{h_m}}{t}$
11:     save result of $P(x_{i' \to i}^t | h_m)$
12: **end for**
13: **for all** $h_m \in H^w$ **do**
14:     calculate $P(h_m | x_{i' \to i}^t) = \frac{P(x_{i' \to i}^t | h_m) P(h_m)}{\sum_{k=1}^n P(x_{i' \to i}^t | h_k) P(h_k)}$
15:     save result of $P(h_m | x_{i' \to i}^t)$
16: **end for**
17: choose the maximum $P(h_m | x_{i' \to i}^t)$ as opponent i' current's preference

---

In Lines 3–6, agent i applies the issue's weighting in each hypothesis h_m to opponent i''s evaluation function $U^{i'}(X^t)$ to estimate the offer's utility. According to the estimation results, agent i updates C_{h_m} (Lines 7–8) and then applies Eq. (8) to calculate $P(x_{i' \to i}^t | h_m)$ for each hypothesis h_m (Lines 9–12). Finally, agent i applies the Bayesian rule (Eq. 7) to calculate the posterior probability of each hypothesis h_m (Lines 13–16) and chooses the hypothesis that has the maximum posterior probability as opponent i' current's preference (Line 17).

2.4 Counter-Offer Proposition

According to opponent i''s concession detail, agent i can estimate its opponent i''s target utility in the next negotiation round and propose a counter-offer whose utility is greater than opponent i''s next target utility.

The concession for each issue in the offer has a limit defined by the maximum and minimum values of the negotiation issue. When a negotiation issue has reached its maximum or minimum value, agent i cannot make any further concession on this negotiation issue.

The concession for each issue also depends on its evaluation function $E_j^i(x_j)$. If the evaluation result of issue j is increasing with the increasing of x_j, the concession on issue j should decrease the issue's value, thus opponent i' can get more utility from this issue. On the contrary, if the evaluation result of issue j is increasing with the decreasing of x_j, the concession on issue j should increase the issue's value.

Before the issue trade-off starts, agent i needs to calculate its next target utility and opponent i''s next target utility $(V_t^{i'})$ based on current negotiation time t. Then the concession starts from the most important issue of opponent i'. The concession for one issue can go for multiple rounds and the concession value in each round can be set by agent i. Therefore, each round's concession for issue j of agent i can be defined as:

$$x_j = \begin{cases} x_j + \delta & \text{if increase } x_j \text{ cause } E_j^i(x_j) \text{ decrease} \\ x_j - \delta & \text{if increase } x_j \text{ cause } E_j^i(x_j) \text{ increase} \end{cases} \tag{9}$$

where δ donates each round's concession value.

After each round of concession, the new offer is evaluated by using both agents' utility functions and then compare with V_t^i and $V_t^{i'}$, respectively. If the utility of the new offer reachs to opponent i''s target utility $V_t^{i'}$, the new offer's utility will be compared with agent i's target utility V_t^i. If the new offer's utility is bigger than agent i's target utility V_t^i, the new offer will be sent to opponent i'. Otherwise, agent i should try to increase the utility of the issue that is least important for opponent i' until the new offer's utility is greater than agent i's target utility V_t^i, then sends the counter-offer to opponent i'.

After each round of concession, agent i should also check whether the concession value has reached the issue's concession limit, which is represented by term $x_{ij(lim)}^t$. If the limit has been reached, then next round of concession should start on the issue that is less important than the current one.

The detail of the counter-offer proposition procedure is described in Algorithm 3 as follows.

In Lines 1–3, agent i first updates itself target utility V_t^i and opponent i''s target $V_t^{i'}$. Then, agent i starts to make concession from the most important issue of opponent i' (Lines 4–12). After the utility concession, if the offer's utility greater then agent i's target utility, agent i will send the new offer to opponent i' (Lines 14–15). Otherwise, agent i will try to gain utility from the least important issue of opponent i' until the offer's utility exceeds agent i's target utility (Lines 17–26).

Algorithm 3 Counter-offer Proposition

1: apply t to concession equation $V_t^i = V_{max}^i - (V_{max}^i - V_{min}^i) * (\frac{t}{t_{max}^i})^\gamma$
2: update agent i's target utility V_t^i
3: update opponent i''s target utility $V_t^{i'}$
4: set issue j as the most important issue of opponent i'
5: **while** $U^{i'}(x_{i'\to i}^t) < V_t^{i'}$ **do**
6:     make concession of issue j by δ
7:     update $x_{i'\to i}^t[j]$
8:     evaluate new offer $U^{i'}(x_{i'\to i}^t)$
9:     **if** issue j reaches concession limit $x_{ij(lim)}$ **then**
10:         find next less important issue according opponent i''s preference
11:         set issue j to this less important issue
12:     **end if**
13: **end while**
14: **if** $U^i(x_{i'\to i}^t) > V_t^i$ **then**
15:     send new offer $x_{i\to i'}^t$
16: **else**
17:     set issue j as the least important issue of opponent i'
18:     **while** $U^i(x_{i'\to i}^t) < V_t^i$ **do**
19:         increase utility of issue j
20:         update $x_{i'\to i}^t[j]$
21:         evaluate new offer $U^i(x_{i'\to i}^t)$
22:         **if** issue j reaches concession limit $X_{ij(lim)}$ **then**
23:             find next important issue according opponent i's preference
24:             set issue j to this issue
25:         **end if**
26:     **end while**
27:     send new offer $x_{i\to i'}^t$
28: **end if**

---

3 Experiment

In this section, we present our experimental results and analyse our negotiation approach's performance. Section 3.1 describes the experimental settings that have been applied in the experiments. Section 3.2 shows the experimental results and performance analysis in three different experimental scenarios as shown in Table 1. Section 3.3 analyses our negotiation approach in a special negotiation case.

Table 1 Experimental scenarios

Scenario	Preference learning	Issue trade-off
1	No agent	No agent
2	Agent 1	Agent 1
3	Agent 1 and 2	Agent 1 and 2

3.1 *Experimental Setting*

In the experiments, our negotiation approach is tested in three different scenarios as shown in Table 1, which are: (1) both agents do not apply the preference learning and the issue trade-off during the negotiation; (2) only one of the negotiation agent applies the preference learning and the issue trade-off; and (3) both agents apply the preference learning and the issue trade-off during the negotiation.

In Scenarios 1 and 2, when an agent does not apply the preference learning and the issue trade-off approaches that described in Algorithms 2 and 3 respectively, it will simply maximise its own utility and do not consider its opponent's utility. More precisely, when such self-interest agent tries to propose offers, it will increase its offer's utility from the most important issue until the issue's limit has been reached, then the utility increasing goes to next important issue.

For each experimental scenario, the negotiation issue's setting and the agent's initial parameters are same. An issue's minimal value is randomly selected from 0 to 500 and the maximum value is randomly selected from 1000 to 2000. Also, the preference values of all five negotiation issues are random numbers between 0 and 1. For an agent, its minimum target utility is randomly selected from 0 to 0.1 and its maximum target utility is randomly selected from 0.9 to 1. The deadlines for both agents are set to 1000 and their concession strategies γ are equal to 1. In order to simplify the experiments, we use the same evaluation function to evaluate all negotiation issues, which can be defined as:

$$E_j(x_j) = \frac{x_j - min_j}{|max_j - min_j|}, \tag{10}$$

where x_j represents the value of issue j and min_j and max_j represent the minimum and the maximum values of issue j, reflectively. The detail of our experiment parameters are shown in Tables 2 and 3.

Table 2 Both agents' parameters

Agent	V_{max}	V_{min}	t_{max}	γ	$E_j(x_j)$
Agent 1 and 2	[0.9, 1]	[0, 0.1]	1000	1	$\frac{x_j - min_j}{|max_j - min_j|}$

Table 3 Negotiation issue's parameters

Issue	max_j	min_j	Preference	$E_j(x_j)$ shape
Issue 1–6	[1000, 2000]	[0, 500]	[0, 1]	{up, down}

3.2 Results and Analysis

For each of the experimental scenario, we have tested the offer's utility when it has been accepted by agent 1 and agent 2. Also, we recorded the time needed by agents to accept an offer in the three scenarios. By comparing the experimental results of the three different scenarios, we can understand the overall performance of the preference learning and issue trade-off algorithms in our negotiation approach. Furthermore, we have tested our negotiation approach in different number of negotiation issues (from two issues to six issues), thus we could have a glimpse of how issue number could affect the performance of our negotiation approach. Since our negotiation approach has employed Bayesian learning to predict the opponent's preference, the prediction results can be greatly affected by the opponent's preference and the acceptance range on the negotiation issues. Such affection could decrease the accuracy of our experimental results. In order to solve this problem, the experiment in each scenario is run for 1000 times and the average results are recorded, thus our experimental results are robustness and generality.

Figure 1 shows the average utility when an offer was accepted by agent 1 in the three experimental scenarios. Figure 2 presents the average utility when agent 2 accepts the offer. Figure 3 shows the total utility of agent 1 and agent 2.

In Fig. 1, we can find that when both agent 1 and agent 2 did not apply our negotiation approach, agent 1's utilities are below 0.55 from 2-issue tests to 6-issue tests. Figure 2 shows the similar result of agent 2's utility tests. Such experimental results are not surprise, since when both self-interest agents try to maximise their own utilities, the negotiation usually will not end with a high utility agreement.

In Scenario 2, after agent 1 has applied our negotiation approach, we can see from Figs. 1 and 2 that not only agent 1's utility has obvious increment, but also agent 2's utility has increased slightly. Such experimental results indicate that although only agent 1 has applied our negotiation approach, but through the preference learning

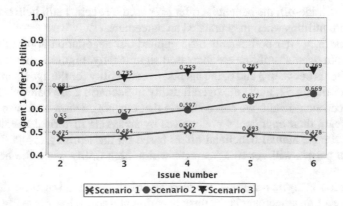

Fig. 1 Agent 1's average utility

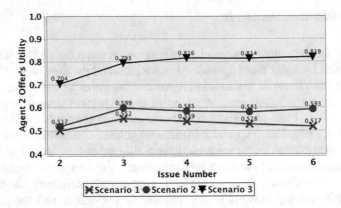

Fig. 2 Agent 2's average utility

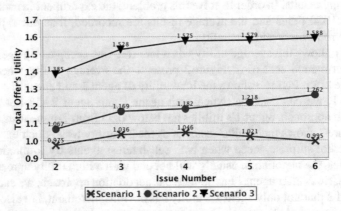

Fig. 3 Agent 1 and 2's total utility

and the issue trade-off, the beneficial offer proposed by agent 1 still help both agents to gain more utilities when they reached an agreement.

In Scenario 3, after both agents have applied our negotiation approach, we can find that both agents' utilities are increased significantly. Figure 3 shows when the issue number is between 3 and 6, the agents' overall utility for the agents in Scenario 3 are higher than 1.52, while the overall utility in Scenario 2 are below 1.26. Based on this experiment result, we can conclude that our negotiation approach can help agents to learn their opponent's preferences by analysing historical counter-offers, and then propose mutual beneficial offers based on the learning result. Thus both negotiation parties will get an agreement with a high utility when the negotiation ends.

In addition to agents utilities, we also record the negotiation time that agents needed to reach an agreement in the three experimental scenarios. From Fig. 4 we can see that agents in Scenario 3 have used least time to complete the negotiation, while agents in the Scenario 1 need most time to reach an agreement. The experimental

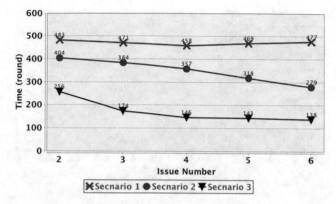

Fig. 4 The time needed when agent 1 and 2 reach an agreement

results indicate that the preference learning in our negotiation approach can help an agent to have better understanding of its opponent's preference and efficiently propose a satisfying offer for its opponent before agents concede too much target utilities.

3.3 Case Study

In this section, we present a case study in detail to analyse how our negotiation approach could affect agents offers in each negotiation round. The negotiation deadlines for both agents are set to the 50th round and all other experimental settings are same as the parameters listed in Table 2. The case study has included four negotiation issues and the detail of each issue's settings are listed in Table 4.

The case study has been conducted over all the three experimental scenarios. During the negotiation, when agent 1 or agent 2 proposed an offer, this offer's utility has been calculated by both agents' utility functions and the results are recorded. The purpose of this experiment is to see how agents utilities will be changed during the whole negotiation in the three different experimental scenarios, thus we may

Table 4 Negotiation issue's parameters

Issue No.	MAX	MIN	Agent 1 preference	Agent 2 preference
1	1745.84	350.42	0.1811	0.3312
2	1891.01	462.28	0.2933	0.4189
3	1746.33	200.50	0.0411	0.1633
4	1244.57	145.75	0.5076	0.0866

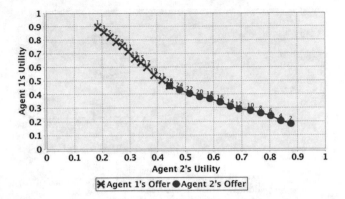

Fig. 5 Agent 1 and 2's utility in Scenario 1

understand what kinds of affections that our negotiation approach can bring to each negotiation agent.

Figures 5, 6 and 7 show the experimental results of the case study. The y-axis in these three charts represents an offer's utility that evaluated by agent 1, while the x-axis represents the offer's utility that evaluated by agent 2. In these charts, we can see two lines, the line with cross points was generated from agent 1's offers and the line with circle points was generated from agent 2's offers. Furthermore, the number that marked on the utility line represents the negotiation round when the offer was generated.

In Fig. 5, we can find that when both agents did not applied our negotiation approach, the utility that agent 1 and agent 2 can gain from a same offer has quite large difference at the beginning of the negotiation. Take the data in the first negotiation round as an example, when agent 1 proposed an offer which has 0.89 utility, agent 2 can only gain 0.185 utility from this offer. Contrary, in the second negotiation round,

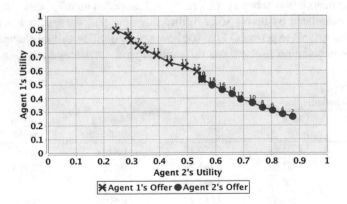

Fig. 6 Agent 1 and 2's utility in Scenario 2

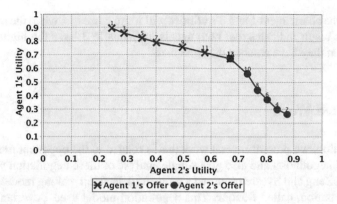

Fig. 7 Agent 1 and 2's utility in Scenario 3

agent 2 proposed an offer that has 0.88 utility, but agent 1 can only gain 0.183 utility from this offer. Such experimental result is not surprised, since at the beginning of the negotiation, agents' target utilities are usually close to 1, thus agents need to try their best to gain utilities from negotiation issues, which leave only little benefit to their opponents. We can see from Fig. 5 that with the negotiation keeps going, both two utility lines are extend themselves toward the same direction (the centre of the chart) and are finally merged with each other at the 26th negotiation round. This is because both agents have decreased their target utilities slightly in each negotiation round. Consequently, both agents can reach their target utilities so as to reach an agreement.

In the Fig. 6, we can see that after agent 1 has applied our negotiation approach, the cross points on the utility line generated by agent 1's offers have the bigger horizontal range comparing with the point's range in the Fig. 5. Obviously, the larger horizontal range between each cross point indicates the utility that agent 2 gets from agent 1's offer has big increment after each negotiation round. Such result is mainly because after agent 1 has studied agent 2's preference, agent 1 can trade off negotiation issues. Therefore, agent 2 can gain more utility from the mutual beneficial offer proposed by agent 1. Through further analysing of Figs. 5 and 6, we can find that there is no significant difference between the utility lines generated by agent 2's offers in the two charts. Such result is quite normal, since in both Scenarios 1 and 2, agent 2 has not applied our negotiation approach, thus agent 2 has only considered to generate offers that could maximise its own utility.

Figure 7 shows the utility change after both agents have applied our negotiation approach. We can see that the vertical range between the circle points has obvious increment compare with the range in Fig. 6. Such change indicates that after the agent 2 has applied our negotiation approach, agent 1 can get more utility from the offer proposed by agent 2 as well. Besides, by comparing the utility line's merge point in the three figures, we can find that the merger point in Fig. 7 is most close to the right up corner of the chart. Apparently, if the merge point is more close to the right up corner of the chart, agent 1 and 2's utilities will more close to 1 when the negotiation ends. This result indicates that both agents in Scenario 3 have the highest overall utility when they have reached the agreement.

4 Related Work

In this section, we present several work that is related to the opponent modelling in automated negotiation and also give simple analyse of these negotiation works.

In [12], Zeng and Sycare proposed a sequential decision making model for multi-issue negotiation, called **Bazaar**. This negotiation model used Bayesian Learning algorithm to learn the opponent's reservation value of certain negotiation issue. During the negotiation, once the agent receives information that comes from its opponent or the outside world, the agent will update its beliefs about the opponent agent's reservation value. Our negotiation approach also employed Bayesian learning algorithm to model the opponent agent. But instead of trying to learn the opponent's reservation value, our negotiation approach focuses on the preference predication, which plays an important role on the proposition of mutual beneficial offers for multi-issue negotiations.

Soo and Hung [11] used Q-Learning algorithm to predict opponent's preferences. During the negotiation, if the opponent has rejected an offer, then this offer is marked as a negative instance for the Q-learning algorithm, while a counter-offer proposed by the opponent gives a positive reward. However, their negotiation approach assumes that the opponent's reservation price is public information, which is very hard or impossible in most automated negotiation scenarios. In our negotiation approach, an agent can learn its opponent's preference without knowing any information about the reservation price.

Luo et al. [10] developed a fuzzy constraint based model for bilateral multi-issue negotiation in a semi-competitive environment. In their framework, the negotiation agent uses prioritised fuzzy constraints to represent trade-offs between the different possible values of the negotiation issues. More precisely, each negotiation agent has a set of fuzzy constraints that are ranked according to priorities. During the negotiation, a buyer agent sends the fuzzy constraint which has the highest priority to an seller agent. Then the seller agent can propose an offer based on the received fuzzy constraint. However, if the seller agent rejects the constraint, the seller needs to partially reveal its preference information, thus the buyer agent knows how to relax its fuzzy constraint. In our negotiation approach, agents negotiate with each other by changing offers, instead of changing constraints. Also, our negotiation approach do not require agents to reveal any information about their preferences.

5 Conclusion and Future Work

In this paper, an automated negotiation approach has been presented to help agents to reach win-win agreements during bilateral multi-issue negotiations. The motivation of this approach is to produce mutual beneficial offers for agents through the preference learning and the issue trade-off. Specifically, a set of hypothesises about the opponent's preference are initialised before negotiation starts, and then the Bayesian learning algorithm is used to analyse the counter-offer proposed by the opponent in each negotiation round and the most suitable hypothesis is chosen to help the agent to generates offers. The proposed negotiation approach has been tested in different scenarios, and the experimental results have proved that our negotiation approach can help agents to reduce the time needed to reach an agreement. Also, agents that have applied our negotiation approach can get more utilities when the negotiation ends.

Our future work will focus on handling negotiations among one-to-many or many-to-many agents and improving our negotiation approach to handle situations that a agent's preference may change dynamically during the negotiation.

References

1. Baarslag, T., Hindriks, K.V.: Accepting optimally in automated negotiation with incomplete information. In: Proceedings of the 12th International Conference on Autonomous Agents and Multi-agent Systems, pp. 715–722 (2013)
2. Bui, H.H., Venkatesh, S., Kieronska, D.: Learning other agents'preferences in multi-agent negotiation using the Bayesian classifier. Inte. J. Coop. Inf. Syst. **8**(04), 275–293 (1999)
3. Coehoorn, R.M., Jennings, N.R.: Learning on opponent's preferences to make effective multi-issue negotiation trade-offs. In: Proceedings of the 6th International Conference on Electronic Commerce, pp. 59–68 (2004)
4. Faratin, P., Sierra, C., Jennings, N.R.: Negotiation decision functions for autonomous agents. Robot. Auton. Syst. **24**(3), 159–182 (1998)
5. Fatima, S.S., Wooldridge, M., Jennings, N.R.: Multi-issue negotiation with deadlines. J. Artif. Intell. Res. **27**, 381–417 (2006)
6. Gwak, J., Sim, K.M.: Bayesian learning based negotiation agents for supporting negotiation with incomplete information. In: Proceedings of the International Multi-conference of Engineers and Computer Scientists, pp. 163–168 (2011)
7. Hindriks, K., Tykhonov, D.: Opponent modelling in automated multi-issue negotiation using bayesian learning. In: Proceedings of the 7th International Joint Conference on Autonomous Agents and Multiagent Systems, vol. 1, pp. 331–338 (2008)
8. Jazayeriy, H., Azmi-Murad, M., Sulaiman, N., Udizir, N.I.: The learning of an opponent's approximate preferences in bilateral automated negotiation. J. Theor. Appl. Electr. Comm. Res. **6**(3), 65–84 (2011)
9. Korb, K.B., Nicholson, A.E.: *Bayesian Artificial Intelligence*. Computer and Information Science. Chapman & Hall/CRC (2004)
10. Luo, X., Jennings, N.R., Shadbolt, N., Leung, H., Lee, J.H.: A fuzzy constraint based model for bilateral, multi-issue negotiations in semi-competitive environments. Artif. Intell. **148**(1), 53–102 (2003)

11. Soo, V.-W., Hung, C.-A.: On-line incremental learning in bilateral multi-issue negotiation. In: Proceedings of the First International Joint Conference on Autonomous Agents and Multiagent Systems, pp. 314–315 (2002)
12. Zeng, D., Sycara, K.: Bayesian learning in negotiation. Int. J. Hum. Comput. Stud. **48**(1), 125–141 (1998)

Automated Negotiating Agent with Strategy Adaptation for Multi-times Negotiations

Katsuhide Fujita

Abstract Bilateral multi-issue closed negotiation is an important class for real-life negotiations. Usually, negotiation problems have constraints such as a complex and unknown opponent's utility in real time, or time discounting. In the class of negotiation with some constraints, the effective automated negotiation agents can adjust their behavior depending on the characteristics of their opponents and negotiation scenarios. Recently, the attention of this study has focused on the interleaving learning with negotiation strategies from the past negotiation sessions. By analyzing the past negotiation sessions, agents can estimate the opponent's utility function based on exchanging bids. In this paper, we propose an automated agent that estimates the opponent's strategies based on the past negotiation sessions. Our agent tries to compromise to the estimated maximum utility of the opponent by the end of the negotiation. In addition, our agent can adjust the speed of compromise by judging the opponent's Thomas-Kilmann Conflict (TKI) Mode and search for the pareto frontier using past negotiation sessions. In the experiments, we demonstrate that our agent won the ANAC-2013 qualifying round regarding as the mean score of all negotiation sessions. We also demonstrate that the proposed agent has better outcomes and greater search technique for the pareto frontier than existing agents.

1 Introduction

Negotiation is an important process in forming alliances and reaching trade agreements. Research in the field of negotiation originates in various disciplines including economics, social science, game theory and artificial intelligence (e.g. [5, 6, 14, 16]). Automated agents can be used side-by-side with a human negotiator embarking on an important negotiation task. They can alleviate some of the effort required of people during negotiations and also assist people that are less qualified in the negotiation

K. Fujita (✉)
Faculty of Engineering, Tokyo University of Agriculture and Technology,
Tokyo 184-8588, Japan
e-mail: katfuji@cc.tuat.ac.jp

© Springer International Publishing Switzerland 2016
N. Fukuta et al. (eds.), *Recent Advances in Agent-based Complex
Automated Negotiation*, Studies in Computational Intelligence 638,
DOI 10.1007/978-3-319-30307-9_2

process. There may even be situations in which automated negotiators can replace the human negotiators. Another possibility is for people to use these agents as a training tool, prior to actually performing the task. Thus, success in developing an automated agent with negotiation capabilities has great advantages and implications.

Motivated by the challenges of bilateral negotiations between automated agents, the automated negotiating agents competition (ANAC) was organized [8, 10, 11, 22]. The purpose of the competition is to facilitate research in the area of bilateral multi-issue closed negotiation. The setup at ANAC is a realistic model including time discounting, closed negotiations, alternative offering protocol, and so on. By analyzing the results of ANAC, the stream of the strategies of automated negotiations and important factors for developing the competition have been shown [1]. Also, some effective automated negotiating agents have been proposed through the competitions [2, 12, 23].

Recently, for automated negotiation agents in bilateral multi-issue closed negotiation, attention has focused on interleaving learning with negotiation strategies from past negotiation sessions. By analyzing the past negotiation sessions, agents can adapt to domains over time and use them to negotiate better with future opponents. However, some outstanding issues regarding them remain, such as effective use of past negotiation sessions. In particular, the way of understanding the opponent's strategy and negotiation scenarios from the past sessions is unclear. In other words, it is still an open and interesting problem to design more efficient automated negotiation strategies against a variety of negotiating opponents in different negotiation domains by utilizing the past negotiation sessions.

In this paper, we propose an adaptive strategy based on the past negotiation sessions by adjusting the speed of compromising depending on the opponent's strategy, automatically. For judging the opponent's strategy, we need to characterize the opponents in terms of some global style, such as negotiation styles or a known conflict-handling style. One important style is the Thomas-Kilmann Conflict Mode Instrument (TKI) [13, 17]. The TKI is designed to measure a person's behavior in a conflict situation based on the concerns of two people appearing to be incompatible. The proposed agent tries to compromise speedily when the opponent is cooperative and passive. By employing this strategy, our agent achieves an agreement in the earlier stage compared with existing negotiating agents. If agents achieve an agreement in the earlier stage, agents can gain more utility because the time-discounted factor decreases the total utility. In addition, our agent has an effective search strategy for finding the pareto optimal bids.

In the experiments, we demonstrate that the proposed agent outperforms the other agents that participated in the qualifying round of ANAC-2013. We also compare the performance of our agent with that of the state-of-the-art negotiation agents. By analyzing the results, it is clear that our agent can obtain higher mean utilities against a variety of opponents in the earlier steps. Additionally, we demonstrate the change of the utility in multi-times negotiation for analyzing the learning strategies.

The remainder of the paper is organized as follows. First, we describe related works. Second, we show the negotiation environments and our proposed agent's basic strategy. Third, we propose a way of adjusting the compromising speed, and a search method for finding pareto optimal bids. Then, we demonstrate the overall results of the qualifying round of ANAC-13 and some experimental analysis. Finally, we present our conclusions.

2 Related Works

This paper focuses on research in the area of bilateral multi-issue closed negotiation, which is an important class of real-life negotiations. Closed negotiation means that opponents do not reveal their preferences to each other. Negotiating agents designed using a heuristic approach require extensive evaluation, typically through simulations and empirical analysis, since it is usually impossible to predict precisely how the system and the constituent agents will behave in a wide variety of circumstances.

Motivated by the challenges of bilateral negotiations between people and automated agents, the automated negotiating agents competition (ANAC) was organized in 2010 [8, 10, 11, 22]. The purpose of the competition is to facilitate research in the area of bilateral multi-issue closed negotiation. The declared goals of the competition are (1) to encourage the design of practical negotiation agents that can proficiently negotiate against unknown opponents and in a variety of circumstances, (2) to provide a benchmark for objectively evaluating different negotiation strategies, (3) to explore different learning and adaptation strategies and opponent models, (4) to collect state-of-the-art negotiating agents and negotiation scenarios, and make them available to the wider research community. The competition was based on the GENIUS environment, which is a General Environment for Negotiation with Intelligent multi-purpose Usage Simulation [15].

By analyzing the results of ANAC, the stream of the strategies of ANAC and important factors for developing the competition have been shown. Baarslag et al. present an in-depth analysis and the key insights gained from ANAC 2011 [1]. This paper mainly analyzes the different strategies using classifications of agents with respect to their concession behavior against a set of standard benchmark strategies and empirical game theory (EGT) to investigate the robustness of the strategies. It also shows that the most adaptive negotiation strategies, while robust across different opponents, are not necessarily the ones that win the competition. Furthermore, our EGT analysis highlights the importance of considering metrics.

Chen and Weiss proposed a negotiation approach called OMAC, which learns an opponent's strategy in order to predict future utilities of counter-offers by means of discrete wavelet decomposition and cubic smoothing splines [3]. They also present a negotiation strategy called EMAR for this kind of environment that relies on a combination of Empirical Mode Decomposition (EMD) and Autoregressive Moving Average (ARMA) [4]. EMAR enables a negotiating agent to acquire an opponent model and to use this model for adjusting its target utility in real time on the basis of an adaptive concession-making mechanism. Hao and Leung proposed a negotiation strategy named ABiNeS, which was introduced for negotiations in complex environments [9]. ABiNeS adjusts the time to stop exploiting the negotiating partner and also employs a reinforcement-learning approach to improve the acceptance probability of its proposals. Williams et al. proposed a novel negotiating agent based on Gaussian Processes in multi-issue automated negotiation against unknown opponents [23]. Fatima et al. focus on the bilateral multi-issue negotiation between self-interested agents in time-limitation settings [7]. By showing the negation model and the optimal procedure for each party, this paper determined equilibria for each procedure for two different information settings. Kawaguchi et al. proposed a strategy for compromising the estimated maximum value based on estimated maximum utility [12]. These papers have been important contributions for bilateral multi-issue closed negotiation; however, they don't deal with multi-times negotiation with learning and reusing the past negotiation sessions.

Recently, some studies have focused on the divided parts of negotiating strategies in the alternative offering protocol: proposals, responses, and opponent modeling. Effective strategies can be achieved by combinations of these strong strategies depending on the opponent's strategies and negotiation environments. Many of the sophisticated agent strategies that currently exist are comprised of a fixed set of modules. Therefore, the studies for proposing the negotiation strategies focusing on the modules are important and influential. Baarslag et al. focus on the acceptance dilemma: accepting the current offer may be suboptimal, as better offers may still be presented [2]. On the other hand, accepting too late may prevent an agreement from being reached, resulting in a break off with no gain for either party. This paper proposed new acceptance conditions and investigated correlations between the properties of the negotiation environment and the efficacy of acceptance conditions.

3 Negotiation Environments

3.1 Bilateral Multi-issue Closed Negotiation

The interaction between negotiating parties is regulated by a *negotiation protocol* that defines the rules of how and when proposals can be exchanged. The competition used the alternating-offers protocol for bilateral negotiation as proposed in [18, 19], in which the negotiating parties exchange offers in turns. The alternating-offers protocol conforms with our criterion to have simple rules. It is widely studied in the literature, both in game-theoretic and heuristic settings of negotiation [5, 6, 14, 16].

For example, *Agents A* and *B* take turns in the negotiation. One of the two agents is picked at random to start. When it is the turn of agent X (X being A or B), that agent is informed about the action taken by the opponent. In negotiation, the two parties take turns in selecting the next negotiation action. The possible actions are:

Accept: This action indicates that the agent accepts the opponent's last bid.
Offer: This action indicates that the agent proposes a new bid.
End Negotiation: This action indicates that the agent terminates the entire negotiation, resulting in the lowest possible score for both agents.

If the action was an *Offer*, agent X is subsequently asked to determine its next action and the turn taking goes to the next round. If it is not an *Offer*, the negotiation has finished. The turn taking stops and the final score (utility of the last bid) is determined for each of the agents, as follows:

- The action of agent X is an Accept. This action is possible only if the opponent actually did a bid. The last bid of the opponent is taken, and the utility of that bid is determined in the utility spaces of agents A and B.
- The action is returned an EndNegotiation. The score of both agents is set to the lowest score.

The parties negotiate over *issues*, and every issue has an associated range of alternatives or *values*. A negotiation outcome consists of a mapping of every issue to a value, and the set Ω of all possible outcomes is called the negotiation *domain*. The domain is common knowledge to the negotiating parties and stays fixed during a single negotiation session. Both parties have certain preferences prescribed by a *preference profile* over Ω. These preferences can be modeled by means of a utility function U that maps a possible outcome $\omega \in \Omega$ to a real-valued number in the range [0, 1]. In contrast to the domain, the preference profile of the players is private information.

A bid is a set of chosen values $v_1 \ldots v_N$ for each of the N issues (I). Each of these values has been assigned an evaluation value $eval(v_i)$ in the utility space. Each issue has been assigned the normalized weight $(w_i, \sum_{i \in I} w_i = 1)$ in the utility space. The utility is the weighted sum of the normalized evaluation values.

The utility function of the bid($\mathbf{v} = (v_1, \ldots, v_N)$) is defined as (1).

$$U(\mathbf{v}) = \sum_{i=1}^{N} w_i \cdot eval(v_i) \tag{1}$$

A negotiation lasts a predefined time in seconds (*deadline*). The time line is normalized, i.e.: time $t \in [0, 1]$, where $t = 0$ represents the start of the negotiation and $t = 1$ represents the deadline. Apart from a deadline, a scenario may also feature discount factors. Discount factors decrease the utility of the bids under negotiation as time passes. Let d in [0, 1] be the discount factor. Let t in [0, 1] be the current normalized time, as defined by the timeline. We compute the discounted utility U_D^t of an outcome ω from the undiscounted utility function U as follows:

$$U_D^t(\omega) = U(\omega) \cdot d^t \tag{2}$$

At $t = 1$, the original utility is multiplied by the discount factor. Furthermore, if $d = 1$, the utility is not affected by time, and such a scenario is considered to be undiscounted.

3.2   Learning from Past Negotiation Sessions

Recently, automated negotiation agents have had the concept introduced that an agent can save and load information for each preference profile. This means that an agent can learn from previous negotiations, against the same opponent or multiple opponents, to improve its competence when having a specific preference profile. By analyzing the past negotiation sessions, agents can estimate the opponent's utility function based on exchanging bids. For example, the bids an opponent proposes many times in the early stage might be the effective bids for the opponents. The last bid proposed by the opponent might be the lowest utility for agreeing with the bid.

The information an agent can save and load for each preference profile and opponent is as follows: Offered bids, received bids,[1] and exchange sequence of the bids. Therefore, we need to predict or analyze the opponent's utility of bids to utilize the past negotiation sessions.

4   Automated Agent Based on Compromise Strategy

This section shows the compromising strategies [12] based on our proposed strategies.

4.1   Opponent Modeling in Basic Strategy

Our agent estimates the alternatives the opponent will offer in the future based on the opponent's offers. In particular, we estimate them using the values mapping the opponent's bids to our own utility function. The agent works at compromising to the estimated optimal agreement point.

Concretely, our behavior is decided based on the following Eqs. (3), (4).

$$emax(t) = \mu(t) + (1 - \mu(t))d(t) \tag{3}$$

$$target(t) = 1 - (1 - emax(t))t^{\alpha} \tag{4}$$

$emax(t)$ means the estimated maximum utility of a bid the opponent will propose in the future. $emax(t)$ is calculated by $\mu(t)$ (the mean of the opponent's offers in our utility space), $d(t)$ (the width of the opponent's offers in our utility space) when the timeline is t. $d(t)$ is calculated based on the deviation. We can see how favorable the opponent's offer is based on the deviation $(d(t))$ and the mean $(\mu(t))$.

If we assume that the opponent's offer is generated based on uniform distribution $[\alpha, \alpha + d(t)]$, the deviation is calculated as follows.

---

[1]Bids don't include the utility information.

Fig. 1 $target(t)$ when $emax(t)$ is $\mu(t) = \frac{1}{10}t$ $d(t) = \frac{2}{5}t^2$

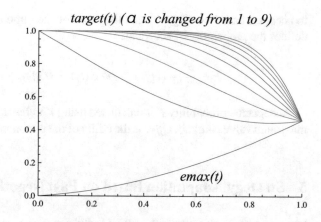

$target(t)$ (α is changed from 1 to 9)

$emax(t)$

$$\sigma^2(t) = \frac{1}{n}\sum_{i=0}^{n} x_i^2 - \mu^2 = \frac{d^2(t)}{12} \tag{5}$$

Therefore, $d(t)$ is defined as follows.

$$d(t) = \sqrt{12}\sigma(t) \tag{6}$$

We consider the means as the weights for the following reason. When the mean of the opponent's action is located at the center of the domain of the utility, $emax(t)$ is the mean plus half of the width of the opponent's offers. However, it is possible to move only in the high direction when the mean of the utility value is low, and the action can be expanded only in the low direction when the mean is high. Therefore, an accurate estimation is made by introducing the weights.

$target(t)$ is a measure of proposing a bid when time is t, and α is a coefficient for adjusting the speed of compromise. It is effective to search for the opponent's utility information by repeating the proposal to each other as long as time allows. On the other hand, our utility value is required to be as high as possible. Our bids are the higher utility for the opponent at the first stage, and approach asymptotically to $emax(t)$ as the number of negotiation rounds increases.

Figure 1 is an example of $target(t)$ when α is changed from 1 to 9. $emax(t)$ is $\mu(t) = \frac{1}{10}t$, $d(t) = \frac{2}{5}t^2$.

4.2 Proposal and Response Opponent's Bids

First, we show the method of selecting the bids from our utility space. Our agent searches for alternatives whose utility is $target(t)$ by changing the starting points randomly by iteratively deepening the depth-first search method. Next, we show the decision of whether to accept the opponent's offer. Our agent judges whether to accept it based on *target*(t) and the mean of the opponent's offers. Equation (7) defines the probability of acceptance.

$$P = \frac{t^5}{5} + (Offer - emax(t)) + (Offer - target(t)) \tag{7}$$

Acceptance probability P is calculated using t, *Offer*, *target*(t) and the estimated maximum value *emax*(t). *Offer* is the utility of the opponent's bid in our utility space.

5 Strategy Adaptation Based on Past Negotiation Sessions

The compromising strategy described in the previous section has following issues:

1. Determination of α adjusting the speed of compromising isn't easy.
2. It doesn't always find the pareto optimal bids in searching bids.

To solve these issues, we propose two strategies using past negotiation sessions.

5.1 Adaptation Strategies Using Past Negotiation Sessions

An opponent's strategy is predictable based on earlier encounters or an experience profile, and can be characterized in terms of some global style, such as the negotiation styles [20, 21], or a known conflict-handling style. One important style is the Thomas-Kilmann Conflict Mode Instrument (TKI) [13, 17]. The TKI is designed to measure a person's behavior in conflict situations. "Conflict situations" are those in which the concerns of two people appear to be incompatible. In this situation, an individual's behavior has two dimensions: (1) assertiveness, the extent to which the person attempts to satisfy his own concerns, and (2) cooperativeness, the extent to which the person attempts to satisfy the other person's concerns. These two basic dimensions of behavior define five different modes for responding to conflict situations: Competing, Accommodating, Avoiding, Collaborating, and Compromising as Fig. 2 shows.

The left side of Table 1 shows the relationships between the condition and cooperativeness, and the right side of Table 1 shows the relationship between the condition and assertiveness. When bid_t (opponent's bid in time t) is higher than μ_h (mean of the bids from past negotiation sessions), our agent regards the opponent as uncooperative. On the other hand, when bid_t is lower than μ_h, our agent regards the opponent as cooperative. In addition, our agent evaluates the assertiveness by comparing between the variance of proposals in the session and that in past negotiation sessions. Usually, assertive agents tend to propose the same bids because they try to push through their proposals by proposing many times. In other words, it is hard for our agent to make win-win agreements when the opponent's bids are disspread. On the other hand,

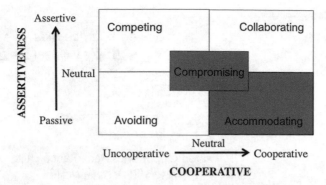

Fig. 2 Overview of Thomas-Kilmann conflict mode instrument (TKI)

Table 1 Estimation of cooperativeness and assertiveness based on past negotiation sessions

Condition	Cooperativeness	Condition	Assertiveness
$u(bid_t) > \mu_h$	Uncooperative	$\sigma^2(t) > \sigma_h^2$	Passive
$u(bid_t) = \mu_h$	Neutral	$\sigma^2(t) = \sigma_h^2$	Neutral
$u(bid_t) < \mu_h$	Cooperative	$\sigma^2(t) < \sigma_h^2$	Assertive

passive agents tend to propose various bids because they change their proposals by searching for win-win agreements. In other words, our agent can make an agreement when the opponent's bids are spread. Considering the above theory, our agent tries to compromise more and more when the opponent is cooperative and passive, which means the opponent is "accommodating" or "compromising" (yellow box in Fig. 2) in the TKI. For judging the opponent's TKI, we employ the past negotiation sessions.

Figure 3 shows the concept of adjusting the speed of compromising in this paper. As Eq. (4) in Sect. 4.1 shows, the speed of compromising is decided by α in $target(t)$. α is set as a higher value at the first stage, and α is decreased when the opponent is "accommodating" or "compromising." By introducing this adjustment algorithm, our agent can adjust its strategy from hardheaded to cooperative more and more when the opponent tries to make agreement. When there is a discount factor, our agent can make an agreement in the early stage by employing the adjustment of α, despite that the existing compromising strategy makes an agreement just before the finish. In addition, our agent can prevent poor compromising because it considers the opponent's strategy and situation.

The detailed algorithm of adapting the agent's strategies based on past negotiation sessions is as follows:

1. Our agent sets α in $target(t)$ to the highest value.
2. It calculates the mean (μ_h) and variance (σ_h^2) of the opponent's bids from past negotiation sessions in appropriate domains.
3. It calculates the utility of offered bid in time t ($u(bid_t)$) and the variance of offered bids from 0 to t ($\sigma^2(t)$).

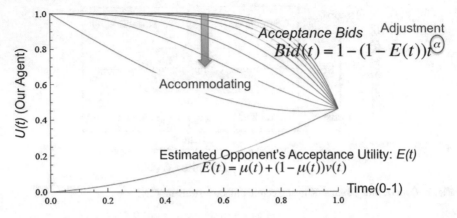

Fig. 3 Adjustment of speed of compromising

4. It compares between μ_h and $u(bid_t)$ to judge the cooperativeness.
5. It compares between σ_h^2 and $\sigma^2(t)$ to judge the assertiveness.
6. It updates the α in $target(t)$ based on the following equation when the opponent is "accommodating" or "compromising":

$$\alpha' = \alpha - \varepsilon \tag{8}$$

(α' is a renewed coefficient for adjusting the speed of compromise, ε is a constant for adjusting the α.)

5.2 Searching for Pareto Optimal Bids

The proposed agent can search for pareto optimal bids based on the similarity between bids. The opponents don't reveal their preferences to each other in the negotiation; therefore, it isn't easy for agents to search for the pareto optimal bids. In this paper, the agent tries to find the bids that are similar to the opponent's first bid because the first bid has high possibility of being the best bid for the opponent.

In this paper, our agent tries to find the most similar bids using the following equation. $\mathbf{v_0}$ means the opponent's bid proposed the first time, and $\mathbf{v_x}$ means the target bid for evaluating the similarity. The similarity between $\mathbf{v_0}$ and $\mathbf{v_x}(sim(\mathbf{v_0}, \mathbf{v_x}))$ is defined as follows:

$$sim(\mathbf{v_0}, \mathbf{v_x}) = \sum_{i=1}^{m} w_i \cdot bool(v_0, v_i) \tag{9}$$

($bool(v_0, v_i)$: if $(v_0 == v_i)$ then return 1 else return 0)

Our agent searches for the bids in which the utility is the same as $target(t)$ and $sim(\mathbf{v}_0, \mathbf{v}_x)$ is highest using the repeated depth-first search algorithm.

6   Experimental Analysis

The performance of our proposed agent is evaluated with GENIUS (General Environment for Negotiation with Intelligent multipurpose Usage Simulation [15]), which is also used as a competition platform for ANAC.

6.1   Qualifying Round Results of ANAC-2013

Nineteen agents were submitted to the competition. The 11 domains were selected from archives submitted by the participants of ANAC-2013. For each pair of agents, under each utility function, we ran a total of 20 negotiations (including the exchange of preference profiles). In other words, 75,240 sessions are run in the qualifying round. The maximum negotiation time of each negotiation session is set to 3 min and normalized into the range of [0, 1]. Table 2 shows mean scores over all the scores achieved by each agent (against all opponents and using all utility functions) and variances.

Note that these means are taken over all negotiations, excluding those in which both agents use the same strategy (i.e. excluding self-play). Therefore, the mean score $U_\Omega(p)$ of agent p in scenario Ω is given formally by:

$$U_\Omega(p) = \frac{\sum_{p' \in P, p \neq p'} U_\Omega(p, p')}{(|P| - 1)} \qquad (10)$$

where P is the set of players and $U_\Omega(p, p')$ is the utility achieved by player p against player p' in scenario Ω. For every domain, due to the normalization of the scores, the lowest possible score is 0 and the highest is 1. The fact that the maximum and minimum scores are not always achieved can be explained by the non-deterministic behavior of the agents: the top-ranking agent on one domain does not always obtain the maximum score on every trial.

As Table 2 shows, our agent has won by a big margin in the qualifying round of ANAC-2013. Considering the variance among the domains, our agent had advantages compared with other agents. Some reasons for this are as follows. First, we try to improve the speed of making agreements by adjusting $emax(t)$. In addition, our agent tries to compromise positively when the opponent is cooperative. Agents couldn't learn from the past negotiation sessions in the past ANAC; therefore, they tried to find effective agreements by eliciting the opponent's utility in the negotiation session. In other words, agents won the utility decreased by the discount factor

Table 2 Results of every combination among ANAC-2013 agents

	Agent	Rank	Mean	Variance
1	**Our Agent**	**1**	**0.562**	**0.00019**
2	Agent Slinkhard	2–3	0.522	0.00132
3	TMFAgent	2–4	0.516	0.00163
4	MetaAgent	3–4	0.495	0.00252
5	GAgent	5–8	0.457	0.00241
6	InoxAgent	5–8	0.455	0.00235
7	SlavaAgent	5–11	0.447	0.00018
8	VAStockMarketAgent	5–11	0.446	0.0052
9	RoOAgent	7–11	0.432	0.00313
10	AgentTalex	7–11	0.431	0.00285
11	AgentMRK2	7–11	0.43	0.00344
12	Elizabeth	12–14	0.387	0.00443
13	ReuthLiron	12–15	0.374	0.00416
14	BOAconstrictorAgent	12–15	0.373	0.00141
15	Pelican	13–18	0.359	0.00434
16	Oriel_Einat_Agent	15–18	0.35	0.00534
17	MasterQiao	15–18	0.345	0.00214
18	Eagent	15–18	0.338	0.00707
19	ClearAgent	19	0.315	0.00109

because they needed to continue many rounds to get enough of the opponent's utility information. However, our agent tries to make agreements in the early stage using the past negotiation sessions when the opponent looks cooperative. Second, our agent could propose pareto optimal bids many times. If agents could offer the pareto optimal bids, the offers are effective and easy for making win-win agreements. Therefore, our agent could find better agreements by the effective search technique.

6.2 Detailed Experimental Analysis

We compare the negotiation efficiency of our proposed agent with eight state-of-the-art negotiation agents that entered the final round of ANAC-2013: GAgent, Meta-Agent, Slava-Agent, TMFAgent, which are implemented by negotiation experts from different research groups.[2] In addition, we added the AgentK, which strategy is the

---

[2]All the agents and the domains that participated in the final round of ANAC-2013 are available in GENIUS 4.2.

Table 3 Mean utility of agreement of each agent against all opponents in different domains

Agent	Acquisition		HouseKeeping	
	Mean	Variance	Mean	Variance
Our Agent	**0.85198**	**0.00681**	**0.47304**	**0.08937**
GAgent	0.90448	0.04179	0.44034	0.061785
Meta-Agent	0.91750	0.05772	0.54888	0.04447
Slava-Agent	0.94569	0.05568	0.41771	0.051393
TMFAgent	0.92361	0.00639	0.43353	0.08918
AgentK	0.87369	0.01248	0.38956	0.047646

Table 4 Mean time of agreement of each agent against all opponents in different domains

Agent	Acquisition		HouseKeeping	
	Mean	Variance	Mean	Variance
Our Agent	**0.33296**	**0.04595**	**0.22151**	**0.17172**
GAgent	0.64804	0.04986	0.50416	0.078815
Meta-Agent	0.84832	0.06358	0.90832	0.073640
Slava-Agent	0.62320	0.04830	0.68524	0.08104
TMFAgent	0.98483	0.00425	0.54105	0.076666
AgentK	0.68889	0.03910	0.62783	0.073640

Table 5 Mean distance to Pareto frontier for each agent against all opponents in different domains

Agent	Acquisition	HouseKeeping
	Mean	Mean
Our Agent	**0.0021133**	**0.13090**
GAgent	0.093006	0.19863
Meta-Agent	0.035986	0.03199
Slava-Agent	0.12254	0.44346
TMFAgent	0.00013182	0.0
AgentK	0.0054941	0.33268

basic compromise strategy [12].[3] For each pair of agents, under each utility function, we ran a total of 20 negotiations (including the exchange of preference profiles).

The negotiation domains can be classified based on the characteristic of weak and strong opposition [1]. Domains with strong opposition mean that the agents have strongly opposite interests over the negotiation outcomes and the gain of one agent must come at the loss of the other agent. Domains with weak opposition refer to those domains in which it is possible for the agents to reach a win-win agreement. In this experiment, we consider two different types of negotiation domains: HouseKeeping

---

[3]For showing the effectiveness of our improvements, AgentK was included in the experiments.

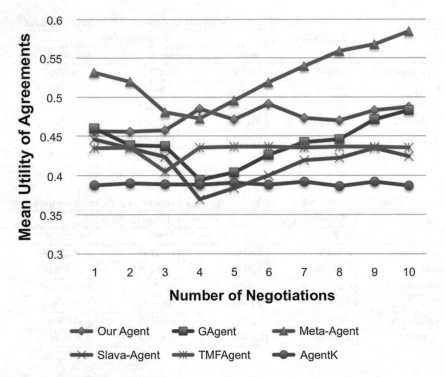

Fig. 4 Changes of mean utility of each agent against all opponents in each round (HouseKeeping)

domain with strong opposition and Acquisition domain with weak opposition. The total number of possible agreements in this domain is 384, and the discount factor and the reservation value of the HouseKeeping domain are set to 0.25 and 0, respectively. Another domain, the Acquisition domain, is much less competitive compared with HouseKeeping. The total number of possible agreements in this domain is 480, and the discount factor and the reservation value are set to 1.0 and 0, respectively. These domains are almost same size; however the opposition and the discount factor between them are totally different.

Table 3 shows the mean normalized scores of agreements for each agent in each domain. Table 4 shows the mean time of agreements for each agent in each domain. Table 5 shows the mean distance to the Pareto frontier for each agent in each domain. The mean distance to the Pareto frontier is defined formally as:

$$\text{meanParetoDistance}(\Omega) = \sum_{\omega \in \text{Offers}} \frac{\min_{\omega_P \in \Omega_P} \text{dist}(\omega, \omega_P)}{|\textit{Offers}|}$$

where $\Omega_P \subset \Omega$ is the set of Pareto efficient possible outcomes, and *Offers* is the set of all bids offered by the agent. The 'dist' function gives the Euclidean distance between two points in the outcome space, defined formally as:

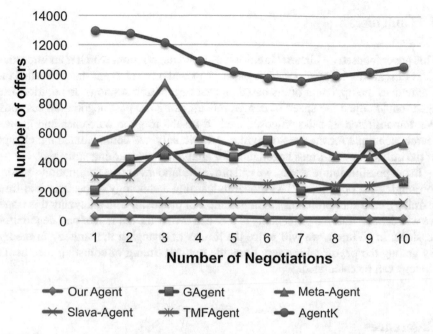

Fig. 5 Changes of mean number of offers of each agent against all opponents in each round (HouseKeeping)

$$\text{dist}(\omega_1, \omega_2) = \sqrt{(U_1(\omega_1) - U_1(\omega_2)) * (U_2(\omega_1) - U_2(\omega_2))}$$

where $U_1(\cdot)$ and $U_2(\cdot)$ give the utilities to players 1 and 2, respectively.

As Table 3 shows, our agent outperforms others expected for Meta-agent with a small variance in the HouseKeeping domain. The main reasons are shown in the results in Tables 4 and 5. Our agent tries to improve the speed of making agreements by adjusting α in $emax(t)$, and compromises positively when the opponent is cooperative. The results of mean time of agreements outperformed compared with other agents in Table 4, definitely. In addition, our agent shows better results than some other agents in Table 5. In other words, these results show the effectiveness of a search method for pareto optimal bids. On the other hand, our agent lost to others in the Acquisition domain because this domain don't have a time-discounting factor.

Figures 4 and 5 show the changes of the mean utility and time of agreements in the HouseKeeping domain. Usually, the results of utility show an inverted parabolic curve in most agents. However, our agent has almost the same utility when the number of negotiations increases. The reasons for this are shown in Fig. 5. Because of the discount factor, the mean utility is influenced a great deal by the mean time of agreement. In fact, the results of mean time show inverted parabolic curves in the all agents except for our agent. The reason for this, shown in Fig. 5, is that we adjust the effective strategy based on TKI. In other words, we can select an effective strategy when the opponent changes its strategies as the negotiation rounds proceed.

7 Conclusion

This paper focused on bilateral multi-issue closed negotiation, which is an important class of real-life negotiations. This paper proposed a novel agent that estimates the alternatives the opponent offers based on past negotiation sessions. In addition, our agent could adjust the speed of compromising using the past negotiation sessions. We demonstrated that the proposed method results in good outcomes and greater search technique for the pareto frontier. Additionally, we demonstrated the change of the utility in multi-times negotiation for analyzing the learning strategies.

In our possible future works, we will prove the amount of past negotiation sessions for judging the opponent's TKI mode. In learning technology (especially real-time learning), cold start problems are important. For proposing and analyzing this issue, we will demonstrate experimentally or prove in theory the amount of past negotiation sessions. In addition, we will prove the timing of changing the strategy in theory. By getting the payoff table every time, the optimal timing of adjusting the agent's strategy can be calculated.

References

1. Baarslag, T., Fujita, K., Gerding, E., Hindriks, K., Ito, T., Jennings, N.R., Jonker, C., Kraus, S., Lin, R., Robu, V., Williams, C.: Evaluating practical negotiating agents: results and analysis of the 2011 international competition. Artif. Intell. J. (AIJ) **198**, 73–103 (2013)
2. Baarslag, T., Hindriks, K.: Accepting optimally in automated negotiation with incomplete information. In: Proceedings of the 2013 International Conference on Autonomous Agents and Multi-agent Systems (AAMAS2013), pp. 715–722 (2013)
3. Chen, S., Weiss, G.: An efficient and adaptive approach to negotiation in complex environments. In: Proceedings of the 19th European Conference on Artificial Intelligence (ECAI-2012), vol. 242, pp. 228–233 (2012)
4. Chen, S., Weiss, G.: An efficient automated negotiation strategy for complex environments. Eng. Appl. Artif. Intell. (2013). doi:10.1016/j.engappai.2013.08.012
5. Faratin, P., Sierra, C., Jennings, N.R.: Using similarity criteria to make issue trade-offs in automated negotiations. Artif. Intell. **142**, 205–237 (2002)
6. Fatima, S.S., Wooldridge, M., Jennings, N.R.: Multi-issue negotiation under time constraints. In: Proceedings of the First International Joint Conference on Autonomous Agents and Multi-agent Systems (AAMAS 2002), pp. 143–150. New York, NY, USA (2002)
7. Fatima, S.S., Wooldridge, M., Jennings, N.R.: Multi-issue negotiation with deadlines. J. Artif. Intell. Res. (JAIR) **27**, 381–417 (2006)
8. Gal, K., Ito, T., Jonker, C., Kraus, S., Hindriks, K., Lin, R., Baarslag, T.: The forth international automated negotiating agents competition (anac2013). http://www.itolab.nitech.ac.jp/ANAC2013/ (2013)
9. Hao, J., Leung, H.F.: Abines: An adaptive bilateral negotiating strategy over multiple items. In: 2012 IEEE/WIC/ACM International Conferences on Intelligent Agent Technology (IAT-2012), vol. 2, pp. 95–102 (2012). doi:10.1109/WI-IAT.2012.72
10. Ito, T., Jonker, C., Kraus, S., Hindriks, K., Fujita, K., Lin, R., Baarslag, T.: The second international automated negotiating agents competition (anac2011). http://www.anac2011.com/ (2011)

11. Jonker, C., Kraus, S., Hindriks, K., Lin, R., Tykhonov, D., Baarslag, T.: The first automated negotiating agents competition (anac2010). http://mmi.tudelft.nl/negotiation/tournament (2010)

12. Kawaguchi, S., Fujita, K., Ito, T.: Compromising strategy based on estimated maximum utility for automated negotiation agents competition (anac-10). In: 24th International Conference on Industrial Engineering and Other Applications of Applied Intelligent Systems (IEA/AIE-2011), pp. 501–510 (2011)

13. Kilmann, R.H., Thomas, K.W.: Developing a forced-choice measure of conflict-handling behavior: the mode instrument. Eng. Appl. Artif. Intell. **37**(2), 309–325 (1977)

14. Kraus, S.: *Strategic Negotiation in Multiagent Environments*. MIT Press (2001)

15. Lin, R., Kraus, S., Baarslag, T., Tykhonov, D., Hindriks, K., Jonker, C.M.: Genius: an integrated environment for supporting the design of generic automated negotiators. Comput. Intell. (2012). doi:10.1111/j.1467-8640.2012.00463.x

16. Osborne, M.J., Rubinstein, A.: *Bargaining and Markets (Economic Theory, Econometrics, and Mathematical Economics)*. Academic Press (1990)

17. Pruitt, D.G.: Trends in the scientific study of negotiation and mediation. Negot. J. **2**(3), 237–244 (1986). doi:10.1111/j.1571-9979.1986.tb00361.x

18. Rubinstein, A.: Perfect equilibrium in a bargaining model. Econometrica **50**(1), 97–109 (1982)

19. Rubinstein, A.: A bargaining model with incomplete information about time preferences. Econometrica **53**(5), 1151–1172 (1985)

20. Shell, G.R.: *Bargaining for Advantage: Negotiation Strategies for Reasonable People*. Penguin Books (2006)

21. Thompson, L.: *Mind and Heart of the Negotiator*, 2nd edn. Prentice Hall Press (2000)

22. Williams, C.R., Robu, V., Gerding, E., Jennings, N.R., Ito, T., Jonker, C., Kraus, S., Hindriks, K., Lin, R., Baarslag, T.: The third automated negotiating agents competition (anac2012). http://anac2012.ecs.soton.ac.uk (2012)

23. Williams, C.R., Robu, V., Gerding, E.H., Jennings, N.R.: Using gaussian processes to optimise concession in complex negotiations against unknown opponents. In: Proceedings of the 22nd International Joint Conference on Artificial Intelligence (IJCAI-2011), pp. 432–438 (2011)

Optimal Non-adaptive Concession Strategies with Incomplete Information

Tim Baarslag, Rafik Hadfi, Koen Hindriks, Takayuki Ito
and Catholijn Jonker

Abstract When two parties conduct a negotiation, they must be willing to make concessions to achieve a mutually acceptable deal, or face the consequences of no agreement. Therefore, negotiators normally make larger concessions as the deadline is closing in. Many time-based concession strategies have already been proposed, but they are typically heuristic in nature, and therefore, it is still unclear what is the right way to concede toward the opponent. Our aim is to construct *optimal* concession strategies against specific classes of acceptance strategies. We apply sequential decision techniques to find analytical solutions that optimize the expected utility of the bidder, given certain strategy sets of the opponent. Our solutions turn out to significantly outperform current state of the art approaches in terms of obtained utility. Our results open the way for a new and general concession strategy that can be combined with various existing learning and accepting techniques to yield a fully-fledged negotiation strategy for the alternating offers setting.

T. Baarslag (✉) · K. Hindriks · C. Jonker
Interactive Intelligence Group, Delft University of Technology,
Mekelweg 4, Delft, The Netherlands
e-mail: T.Baarslag@tudelft.nl; T.Baarslag@gmail.com

K. Hindriks
e-mail: K.V.Hindriks@tudelft.nl

C. Jonker
e-mail: C.M.Jonker@tudelft.nl

R. Hadfi · T. Ito
Department of Computer Science and Engineering,
Graduate School of Engineering, Nagoya Institute of Technology,
Gokiso, Showa-ku, Nagoya 466-8555, Japan
e-mail: rafik@itolab.nitech.ac.jp

T. Ito
e-mail: ito.takayuki@nitech.ac.jp

© Springer International Publishing Switzerland 2016
N. Fukuta et al. (eds.), *Recent Advances in Agent-based Complex
Automated Negotiation*, Studies in Computational Intelligence 638,
DOI 10.1007/978-3-319-30307-9_3

39

1 Introduction

A key insight of negotiation research is that making concessions is crucial to conducting a successful negotiation. There are important reasons to make concessions during the negotiation [25]: it is often used to elicit cooperation from the other, in the hope that the other will reciprocate in kind. Second, it conveys information to the opponent, both about the negotiator's preferences and about the perceptions of the opponent. But most importantly, it is the time pressure of the negotiation itself (typically in the form a deadline or a perceived maximum number of bidding rounds) that operates as a force on the parties to concede [8]. An approaching deadline puts important pressure on the parties to reduce their aspirations, especially when the time pressure heightens, which is referred to as the "eleventh hour effect".

Given the paramount importance of time in bargaining, it is not surprising that many negotiating agents adjust their level of aspiration based on the time that is left in the negotiation. There is a clear rationale behind the design of such agents, given their aim to maximize the chance of reaching an agreement in a limited amount of time. For example, well-known time dependent tactics (TDT's) [10, 11], such as *Boulware* and *Conceder*, are characterized by the fact that they consistently concede throughout the negotiation process as a function of time. Time-based concession curves can also be observed in practice in the Automated Negotiating Agents Competition (ANAC) [3, 32]. However, in the TDT's, as well as in some very effective agents, such *Agent K* [16] (winner of ANAC 2010) and *HardHeaded* [30] (winner of ANAC 2011), the specific concession curve is selected rather arbitrarily, and is not informed by any other insights; therefore, they make largely unfounded choices on how much to concede at each time interval.

Alternatively, behavior dependent tactics (e.g. reciprocating the opponent's concessions by tit for tat [5, 10]) base their decision to make concessions on the actions of the other negotiating party. However, such *adaptive* approaches do not give us any information on how to concede based on time alone.

Work that presents optimal choices of how much to concede includes game theoretic work (e.g. [26]) and single-shot bargaining, also known as the *ultimatum game* [23]. However, these approaches usually assume a complete information setting, or a game where the deal is struck immediately, which we cannot apply to a typical concession-based negotiation. Furthermore, this type of work typically revolves around equilibrium strategies, which assumes full rationality on the part of both agents. We are more interested in optimal solutions for one negotiating party, playing against various classes of acceptance strategies.

This paper aims to find out how time pressure alone, in the form of a deadline, should influence the concession behavior of a negotiator against specific opponent classes. To do so, we employ methods from sequential decision theory to devise negotiation strategies that make *optimal* concessions at each negotiation round. Finally, we show that an agent making these optimal concessions performs better than any other in our experimental setup.

We begin with an example in Sect. 2 that sets the stage for our time-based concession model in Sect. 3. We apply our methods to find optimal concessions against opponents that accept according to acceptance thresholds in Sects. 4 and 5. We subsequently compare the optimal bidding strategy with state of the art bidding strategies in a series of tests (Sect. 5). We conclude our paper with a discussion of related work (Sect. 7) and the contributions of this paper, as well as its implications (Sect. 6).

2 An Example

The following example serves as an illustration of the basic insights that have motivated our approach.

Suppose agent B is negotiating the purchase of a house with a buyer, agent A. As in the rest of this paper, we will only focus on one of the two parties; in this case B.

B has set the opening price at \$300,000, but is (secretly) willing to go down to \$250,000 if necessary. B has to strike a balance between the probability that A buys the house, and getting as much utility out of the agreement as possible (i.e., not conceding too much). Our main question is: what offers or concessions should B make towards A in order to perform optimally and maximize his own outcome? That is, what is the right way to lower the price of the house depending on the remaining time and A's acceptance policy?

Let us consider the easiest case where B has only *one* offer to make to A, which A can either accept or reject. If B makes an offer $x \in [\$250,000; \$300,000]$ to A, then this will yield him the following utility:

$$U(x) = \begin{cases} \frac{x - \$250,000}{\$50,000}, & \text{if } A \text{ decides to accept,} \\ 0, & \text{when } A \text{ rejects the offer.} \end{cases}$$

Now suppose A is prepared to pay up to \$280,000 for the house. Of course, B does not know this, but instead presumes that A's *acceptance threshold* could be anywhere between \$200,000 and \$300,000 with equal probability. It follows that B believes the chance that A accepts decreases linearly in terms of the price offered. Then, B's strategy is simple: he should set the price to

$$\arg\max_x U(x) \cdot P(A \text{ accepts } x), \tag{1}$$

which is equal to

$$\arg\max_x \left(\frac{x - \$250,000}{\$50,000} \right) \cdot \left(1 - \frac{x - \$200,000}{\$100,000} \right). \tag{2}$$

We can readily compute that the maximum is reached for $x = \$275,000$, and so with only one offer to make, B should pick this price in order to optimize his utility. Note how B's offer falls exactly in between his reservation price and his opening price. Fortunately for both, the price x is also actually lower than the maximum A was willing to pay, and therefore, she will buy the house.

This simple case serves as a good intuition to proceed to the more general setting we consider below.

3   Negotiation Model

Our negotiation model builds upon the alternating offers protocol [26]: two agents A and B exchange offers in turns on a fixed negotiation domain Ω that contains all possible negotiation outcomes. A and B have utility functions $U_A : \Omega \to [0, 1]$ and $U_B : \Omega \to [0, 1]$ and reservation values $rv_A, rv_B \in [0, 1]$ respectively. The agents will only propose offers that they deem acceptable, namely bids with higher utility than their reservation value. Likewise, the reservation values acts as the lowest utility they are willing to accept.

The process is illustrated in Fig. 1. With n rounds remaining, B starts by making an offer $B_n \in \Omega$, which A can either respond to with an accept or a counteroffer. If A accepts, the negotiation ends in agreement, and both sides obtain their respective utility. If A rejects the offer, she replies with a counteroffer A_n, the number of remaining rounds decreases by one, and B can make a second offer, B_{n-1}. This process continues until B makes his last offer B_1. If A also turns down this last offer, A and B receive their respective reservation values rv_A and rv_B.

Throughout this work, we are concerned with finding optimal concessions for one of the parties, given certain strategies used by the opponent. We focus on agent B for his *Bidding* role, and on A for her *Accepting* role. In particular, we do not focus on what offers A should generate, or what acceptance policy B should employ.

There is an important class of strategies that allows us to make some further simplifications. In line with our aim to find concession curves that depend only on the time remaining, we will assume B's strategy is not adaptive (i.e., B does not change his behavior according to the bids that have been exchanged). This means B

Fig. 1 Sequential diagram of the negotiation process. B starts the negotiation with bid B_n, and after every proposal, the other responds with a counteroffer or an Accept

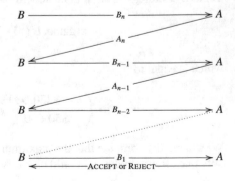

can completely ignore A's bids, and we can code them by ACCEPT or REJECT instead. Other examples of such agents include the family of TDT's [10], the *sit-and-wait* agent [2], and IAMcrazyHaggler [31]. Note that this then effectively defines an one-sided bidding protocol in which B submits n bids to A in a sequential manner, which is equivalent to a repeated ultimatum game with reservation values [28].

The fact that B does not adapt to his opponent puts him in a significantly more difficult epistemic position than in a typical negotiation, since B could normally gather valuable information from A's counter offers. This holds in a very fundamental sense: apart from the incoming offers, B can only distinguish n different possible states, namely the number of rejects that he received. This means that we can already compute the appropriate *concession curve* before the start of the negotiation; i.e., the bids or utilities that need to be sent out at every round.

Also note that as time moves *forward*, the indexing of the offers by B runs *backward*. This has the advantage that it simplifies the calculations of certain recurrence relations we will encounter in this paper, as the expected utility with $j + 1$ rounds remaining depends on the expected utility in the future (i.e., with j rounds remaining). Secondly, it allows us to define our model without a preset deadline, as we do not need to specify the maximum number of rounds beforehand.

4 Conceding and Accepting

The aim of this paper is to find the right concession behavior for B given that he only has n offers to try out. B will only propose offers that he himself deems acceptable, namely bids in the following set:

$$O_B = \{\omega \in \Omega \mid U_B(\omega) \geq rv_B\}. \tag{3}$$

B can make his bids in many ways (see also our discussion of related work in Sect. 7).

A well-known method using concession curves are the TDT's such as *Boulware* or *Conceder*. When there are j rounds remaining (out of a total of n rounds), this strategy makes a bid with utility closest to

$$P_{min} + (P_{max} - P_{min}) \cdot \left(1 - k - (1 - k) \cdot \left(1 - \frac{j}{n}\right)^{\frac{1}{e}}\right),$$

for certain choices of k, e, P_{min} and P_{max}, which control the concession rate and the minimum and maximum target utilities. These tactics form the basis of many successful negotiation strategies [3, 11, 16, 30], in which variants of the above formula are used to decide the appropriate concessions, combined with advanced techniques such as preference modeling and strategy prediction. In what follows, we will propose a different kind of concession curve that is also able to provide the groundwork for designing an advanced negotiation agent.

B's optimal strategy of course depends on how A chooses to accept or reject. While B makes his offers, A has to solve some kind of stopping problem to decide when the offer is sufficient.

Player A will have to either accept or reject a bid $\omega \in \Omega$ in every round j. A may update her beliefs in round j based on ω and the bidding history of the opponent, which is an element h_j from the set of offers $\mathcal{H}_j = \Omega^{n-j}$ made by B. Note that we only need to consider the bids made by B, since the rejects of A are already implicitly represented. For example, for $n = 8$ and $j = 5$, the information set of A is a history of rejected offers $h_5 = (B_6, B_7, B_8) \in \mathcal{H}_5$, adjoined with the bid B_5 of the current round.

Hence, we can represent A's acceptance strategy in round j as a function

$$\alpha : \mathcal{H}_j \times \Omega \to \{\text{ACCEPT}, \text{REJECT}\}. \tag{4}$$

In this paper, we will focus on a specific set of strategies that accept according to *acceptance thresholds*. That is, A's acceptance strategy in round j is specified by a utility constant α_j (possibly dependent on rv_A) such that

$$\alpha(h_j, \omega) = \begin{cases} \text{ACCEPT if } U_A(\omega) \geq \alpha_j, \\ \text{REJECT \ otherwise.} \end{cases} \tag{5}$$

We believe acceptance thresholds are a natural set of acceptance strategies to consider, since it is reasonable to assume that if A finds a bid acceptable, then so is any bid with higher utility.

One of the simplest acceptance strategies for A is *satisficing*: accepting any offer with a utility above her reservation value by setting all α_j equal to rv_A. This is done, for example, when a negotiator is more concerned with getting any deal at all than reaching the best possible deal (e.g., [22]). Needless to say, this is a very simple acceptance strategy, as A normally wants to get as much out of the negotiation as possible. In other applications, illustrated in Fig. 2, A's threshold would be higher at the beginning of the negotiation, and would slowly decrease towards rv_A. However, as we shall see in the next section, the case of a satisficing acceptor already requires highly effective conceding behavior by B.

A might also employ a more fundamental approach by trying to optimize her own utility, taking into account the number of rounds remaining. Optimal stopping theory provides optimal solutions for A for the case that A has incomplete information about B's offers. From [7, 19, 33], we know that the optimal solution against a bidder B with completely unknown utility goals is as follows:

Proposition 1 *When B makes random bids of utility uniformly distributed in $[0, 1]$, and with j offers still to be observed, A's optimal acceptance strategy is to accept an offer of utility x exactly when $x \geq v_j$, where v_j satisfies the following equation:*

$$\begin{cases} v_0 = \text{rv}_A, \\ v_j = \frac{1}{2} + \frac{1}{2}v_{j-1}^2. \end{cases} \tag{6}$$

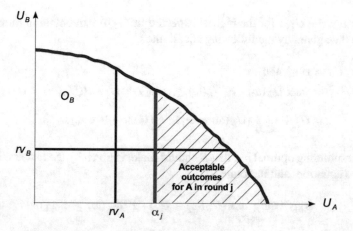

Fig. 2 A's acceptable offers for round j consist of all bids ω with $U_A(\omega) \geq \text{rv}_A$, while B's possible offers O_B consist of all bids ω such that $U_B(\omega) \geq \text{rv}_B$

As it will turn out, this strategy is closely related to the strategy B should follow against satisfying acceptors.

5 Making Optimal Offers

We will now outline our general method to make optimal offers, which extends our running example from Sect. 2 to a general domain with an arbitrary number of remaining rounds.

Suppose A uses an acceptance strategy based on acceptance thresholds $(\alpha_0, \alpha_1, \ldots)$, with $\alpha_j \geq \text{rv}_A$; that is, A accepts an offer $\omega \in \Omega$ with j remaining rounds exactly when $U_A(\omega) \geq \alpha_j$. Suppose we have $j + 1$ rounds to go, and B has to decide the optimal bid B_{j+1} to make.

The general formula for the expected utility $U_{j+1}(\omega)$ for B of offering $\omega \in O_B$ is as follows:

$$U_{j+1}(\omega) = \begin{cases} U_B(\omega), & \text{if } A \text{ accepts,} \\ \text{the expected utility} \\ \text{with } j \text{ remaining rounds, if } A \text{ rejects.} \end{cases}$$

When there are no more rounds remaining, B will get rv_B, hence $U_0(\omega) = \text{rv}_B$. The recursive nature of this equation allows us to employ techniques from sequential decision theory to formulate the optimal way to make concessions.

We will write U_{j+1} for the highest expected utility B can obtain in round $j + 1$, which is thus given by the following equations:

$$U_0 = \mathrm{rv}_B, \text{ and}$$

$$U_{j+1} = \max_{\omega \in O_B} U_B(\omega) \cdot P(U_A(\omega) \geq \alpha_{j+1}) + U_j \cdot P(U_A(\omega) < \alpha_{j+1})$$

$$= U_j + \max_{\omega \in O_B} (U_B(\omega) - U_j) \cdot P(U_A(\omega) \geq \alpha_{j+1}).$$

The corresponding optimal *bid* that B should make is given by the ω that maximizes the above equation, and therefore,

$$B_{j+1} = \arg\max_{\omega \in O_B} (U_B(\omega) - U_j) \cdot P(U_A(\omega) \geq \alpha_{j+1}).$$

Solving this equation would be straightforward, if it were not for the fact that B in general does not have full knowledge of a number of aspects in this equation: U_A is of course unknown to B, and so are the acceptance thresholds α_j. To make matters worse, the acceptance thresholds generally depend on the reservation value of A, which is also unknown to B.

B will therefore have to make some assumptions about U_A and her concession thresholds. Assume that B has an estimation of the reservation value of A, which does not change according to A's behavior. As in [20], the estimation is characterized by a probability distribution $F_j(x)$ for every remaining round j, where $F_j(x)$ denotes the probability that A's reservation value is no greater than x.

We will now analytically solve the specific case of an opponent with a satisficing accepting strategy. To get concrete examples of concession curves (see Fig. 3), we will consider a classical buyer-seller scenario as in [10, 11] to evaluate our solutions. We study the negotiation scenario of *Split the Pie* [26, 29], where two players have to reach an agreement $x \in [0, 1]$ on the partition of a pie of size 1. The pie will be partitioned only after the players reach an agreement. In this setting, we instantiate $\Omega = [0, 1]$ to represent a pie of size 1, with A and B having opposing preferences on them: $U_B(x) = x$ and $U_A(x) = 1 - x$.

We assume that A's acceptance strategy is limited to the class where she accepts any offer that is better than her reservation value; i.e., $\forall_j \alpha_j = \mathrm{rv}_A$, where we assume rv_A is uniformly distributed. The probability that A accepts in round j is now:

$$P(U_A(\omega) \geq \alpha_j) = P(U_A(\omega) \geq \mathrm{rv}_A) = F_j(U_A(\omega)). \tag{7}$$

The general formula now simplifies to

$$U_{j+1} = U_j + \max_{x \geq \mathrm{rv}_B} (x - U_j) \cdot (1 - x). \tag{8}$$

Note that the following holds, even for the general setting:

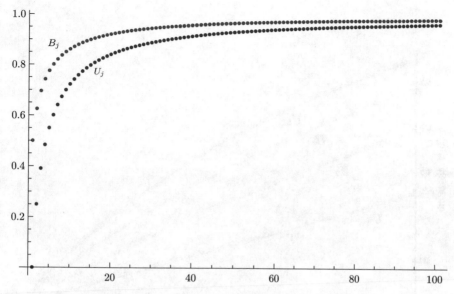

Fig. 3 Graph of B_j and U_j for remaining rounds $j \in \{0, 100\}$

Proposition 2 *When B has more rounds remaining, B can expect to get more utility out of our negotiation. That is, for every remaining round j, we have $U_{j+1} \geq U_j$.*

Proof B can always make the bid $\omega_{\max} = \arg \max_\omega U_B(\omega)$ with $j + 1$ rounds remaining, to get at least as much utility as in the next round. □

With the help of Proposition 2, we can show the maximum in Eq. (8) is attained for $x = B_j$, where B_j satisfies the following relationship:

$$\begin{cases} B_1 = \frac{rv_B + 1}{2}, \\ B_{j+1} = \frac{1 + U_j}{2}. \end{cases} \tag{9}$$

This yields the following recurrence relation for U_j:

$$\begin{cases} U_0 = rv_B, \\ U_{j+1} = \frac{1}{4}(U_j + 1)^2. \end{cases} \tag{10}$$

With these equations, we can now compute the expected value of the optimal bids B has to make in terms of his own reservation value (see Fig. 4). For example:

$$U_1 = \frac{1}{4}(1 + rv_B)^2,$$

$$U_2 = \frac{1}{64}(5 + rv_B(2 + rv_B))^2,$$

$$\vdots$$

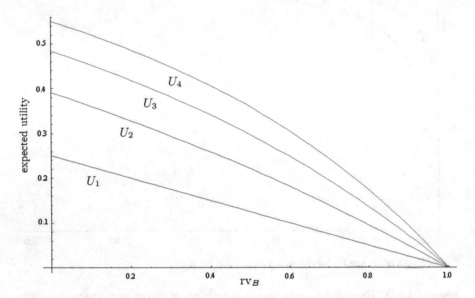

Fig. 4 Expected utility U_j for B with $j \in \{1, 2, 3, 4\}$ rounds to go, depending on B's reservation value rv_B

The corresponding optimal concessions by B are as follows:

$$B_1 = \frac{1 + rv_B}{2},$$

$$B_2 = \frac{1}{8} \cdot (5 + rv_B \cdot (rv_B + 2)),$$

$$\vdots$$

Note that Eq. 10 is related to the logistic map: if we substitute $U_j = 1 - 4x_j$ in Eq. 10 we get the equivalent relation of the logistic map $x_j = x_{j-1}(1 - x_{j-1})$ at $r = 1$, so we cannot expect to solve the recurrence relation. Note, however, that both Eqs. (9) and (10) are expressed in terms of the value U_j. We obtain a much more elegant formulation when we express them in terms of B_j:

Proposition 3 *For all $j \geq 1$,*

$$B_{j+1} = \frac{1}{2} + \frac{1}{2}B_j^2,$$

and

$$U_j = B_j^2.$$

Proof Both statements follow from rewriting Eqs. (9) and (10).   □

Corollary 1 *For* $rv_B = 0$, *there exists the following connection between* B_j, U_j, *and the optimal stopping cut-off values* v_j *from Eq. (6):*

$$B_j = v_j,$$

and

$$U_j = v_j^2.$$

Proof When $rv_B = 0$, then $B_1 = \frac{1}{2}$, and consequently, the B_j sequence has the same starting value and definition as the optimal stopping sequence v_j.   □

Corollary 1 shows that optimal *bidding* against a satisficing acceptor with unknown reservation value is the mirrored version of optimal *accepting* against a bidder that makes unknown offers. In both cases, the idea is the same, but with switched roles: both the optimal bidder and the optimal stopper aim to pick the optimal utility threshold that simultaneously maximizes the expected utility of an agreement and the chance of acceptance, given stochastic behavior by the opponent.

We conclude this section with an example that relates our results to the housing example of Sect. 2.

Example 1 Our assumptions in this section are consistent with our housing example when we scale the pie Ω to [\$250,000; \$300,000], and when $rv_B = 0$ corresponds to the utility of selling the house for \$250,000. Our results indicate that the seller of the house should act as follows: we see that indeed, $B_1 = \frac{1}{2}$, so with one bid remaining, B should make a bid halfway between \$250,000 and \$300,000, as we had calculated before, with an expected utility of $U_1 = \frac{1}{4}$.

Also, $B_2 = \frac{5}{8}$, which means with *two* bids remaining, B should offer \$250, 000 + $\frac{5}{8} \cdot$ \$50,000 = \$281,250. See Table 1 for other entries. Note that since B does not learn anything about A, B actually overestimates the expected utility, and more so when the number of rounds increases (as can be seen in Table 1). This is because B is not able to deduce, from A's rejects, that high utility values for B are simply not attainable.

Table 1 The optimal offers to make and their expected utility in the housing example, given the remaining amount of j bidding rounds

j	Offer	Utility
1	\$275,000	0.25
2	\$281,250	0.39
3	\$284,766	0.48
4	\$287,086	0.55
5	\$288,754	0.60
10	\$293,055	0.74
100	\$299,060	0.96

6 Experiments

In order to test the efficacy of the optimal bidding technique given by Proposition 3, we integrated it into a fully functional negotiating agent. Given j remaining rounds it makes an offer with utility target B_j as defined by Eq. (9). It does not accept any offers and it does not model the opponent in any way.

For the opponents (side A), we selected various well-known negotiation agents available for our setting, including the top three agents of ANAC 2012 [32], namely *CUHK Agent* [15], *AgentLG*, and *OMAC Agent* [9]. We also included the time dependent tactics (TDT's) *Boulware* (with concession factor $e = 0.2$), and *Conceder* ($e = 2$) taken from [10]. As a baseline, we included the *Random Walker* strategy [14], which generates random bids. We then compared the optimal bidder's performance with the same set of strategies on side B. To analyze the performance of different agents, we employed GENIUS [21], which is an environment to design and evaluate automated negotiators' strategies and their components [4].

Note that some of the agents in our setup are originally designed to work with the alternating offers protocol, while we essentially employ a one-sided bidding protocol; however, since our model is a simplified alternating offers protocol, it is easy to adjust the agents to work in our setting: for the opponents, we ignore any bids that are sent out; i.e., when side A accepts, the negotiation ends in agreement, while A's counter-offers count as rejects and are ignored. In effect, this means only A's acceptance mechanisms [6] are used.

For our negotiation scenario we use a discretized version of *Split the Pie*. We set $rv_B = 0$, and we selected varying reservation values for A: $rv_A \in \{0, 0.1, 0.2, \ldots, 0.9\}$. We ran our experiments using varying total number of rounds $n \in \{1, 20, 40, \ldots, 100\}$, and we repeated every negotiation session 5 times for statistical significance.

The results of our experiment are plotted in Fig. 5, with the total amount of rounds n varying between 1 and 100. As is evident from the results, the optimal bidder significantly outperforms all agents in all cases (*one-tailed t-test*, $p < 0.01$ for every n). The good relative performance of the optimal bidder is even more pronounced for $n = 1$, and it is easy to see why. Many of the other strategies will persist in aiming for a high utility, even with only few bids to send out. This results in many break-offs, while optimal bidder will settle for a much lower value in the last rounds with more potential agreements. For example, for $n = 1$, optimal bidder sets his utility target halfway between his reservation value and the maximum attainable; i.e., $B_1 = \frac{rv_B + 1}{2}$. On the other hand, with more bids remaining, optimal bidder acts like an extreme *Boulware* strategy, trying to get as much out of the negotiation as possible. Indeed, optimal bidder and *Boulware* tend to act more similar as n increases. *CUHK Agent* (the winner of ANAC 2012) and *AgentLG* obtain particularly low scores. The main reason for this is that these strategies are very behavior-dependent and do not take into account the remaining time as much as they need to.

Note that almost all agents obtain higher utilities with more negotiation rounds. This is to be expected, as more time allows for a more fine-grained search of what

utility

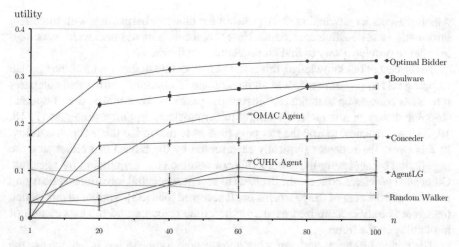

Fig. 5 The utility obtained by the bidding strategies in our experiments for different values of the total number of rounds n. The vertical bars indicate one standard deviation from the mean.

is acceptable for the opponent. The only exception is *Random Walker*, who only increases the chances to make a disadvantageous bid for itself with more available time.

7 Related Work

Concession making is an important process as it might critically affect the outcomes of the negotiation. Despite its importance, the literature lacks a fully comprehensible theory where the uncertainty settings as well as the optimal concession strategies are formally represented.

Our interest lie in investigating the optimal concession strategies against particular classes of acceptance strategies. The concessions we are referring to, relate to the subjective type of concessions as proposed in the categorization of [17]. However, our focus is on the type of concessions that are mainly driven by time dependence for the offers generation, such as the time dependent tactics (*Boulware*, *Conceder*) found in [10–13], or the *monotonic concession protocol* [24], where each agent first starts with the best offer for itself and then iteratively makes compromises to the other agent.

The bargaining game we take as main setting could be seen as an instance of the ultimatum game, in which a player proposes a deal that the other player may only accept or refuse [28]. Another work, [27], explores a similar bargaining model as well; that is, models with one-sided incomplete information and one sided offers. It investigates the role of confrontation in negotiations and uses optimal stopping is to decide whether or not to invoke conflict. Our setting can also be found in [1], which presents an alternating offer protocol for bilateral bargaining with imperfect information and deadline constraints. Here too, the authors use backward induction, in order to provide a way to find all sequential equilibria.

To the best of our knowledge, this is the first work that makes usage of the optimal stopping rule to *generate* offers in an incomplete information setting and compares it to other concession techniques. Most of the previous work makes use of optimal stopping theory in a negotiation setting to formulate acceptance strategies [7, 18, 19, 33]; for instance, to find the stopping time as to maximize the acceptor's utility. In this work, the acceptor (typically represented by the buyer) can either stop the negotiation by accepting the seller's offer or continue the negotiation by rejecting. Other such problems concern the decision to accept sequential job offers while trying to maximize the sum of the payments of all accepted jobs [33]. Similar to our model, the agent is only incentivized by the limited time resource and the expectation of high utility in the future.

The major difference with our work and the above approaches is that we use the optimal stopping rule for concessions, instead of focusing on the optimal time to accept. Our work is defined more as the *complimentary* version of these approaches, in the sense that our formulation of optimal bidding rules happen to resemble optimal acceptance rules. Another key point is that we do not assume that the players' strategies are fixed, which allows us to formulate optimal bidding strategies against certain types of accepting strategies.

8 Conclusion

This paper presents a theoretical model to formulate optimal concession curves against strategies that decide when to accept using acceptance thresholds. We calculated these optimal concessions by employing sequential decision methods, and we showed that they significantly outperform state of the art concession strategies, even against a much wider set of acceptance strategies. As far as the authors are aware, this is the first time such informed time-based concession strategies have been formulated and tested in practice.

Our results demonstrate that our optimal bidding mechanism is an effective way for a negotiating agent to take into account the passing of time. We believe even more effective concession curves can be computed by studying more general negotiation settings (e.g., dynamic, multi-lateral, and concurrent negotiation environments) and broader ranges of acceptance strategies than we did in this paper. Further improvement could be made by introducing a form of learning to the optimal bidder. Eventually, we envision a design of an automated negotiator that incorporates our optimal concession curve with regard to time-related concessions, while other types of concessions (e.g., to elicit cooperation or to convey information) are handled separately by other concession modules.

References

1. An, B., Gatti, N., Lesser, V.: Bilateral bargaining with one-sided uncertain reserve prices. Auton. Agents Multi-Agent Syst. **26**(3), 420–455 (2013)
2. An, B., Sim, K.M., Tang, L.G., Miao, C.Y., Shen, Z.Q., Cheng, D.J.: Negotiation agents' decision making using markov chains. In: Ito, T., Hattori, H., Zhang, M., Matsuo, T. (eds.) Rational. Robust, and Secure Negotiations in Multi-Agent Systems, volume 89 of Studies in Computational Intelligence, pp. 3–23. Springer, Berlin (2008)
3. Baarslag, T., Fujita, K., Gerding, E.H., Hindriks, K., Ito, Takayuki, Jennings, Nicholas R., Jonker, Catholijn, Kraus, Sarit, Lin, Raz, Robu, Valentin, Williams, Colin R.: Evaluating practical negotiating agents: results and analysis of the 2011 international competition. Artif. Intell. **198**, 73–103 (2013)
4. Baarslag, T., Hindriks, K., Hendrikx, M., Dirkzwager, A., Jonker, Catholijn: Decoupling negotiating agents to explore the space of negotiation strategies. In: Marsa-Maestre, I., Lopez-Carmona, M.A., Ito, T., Zhang, M., Bai, Quan, Fujita, Katsuhide (eds.) Novel Insights in Agent-based Complex Automated Negotiation. Studies in Computational Intelligence, vol. 535, pp. 61–83. Springer, Japan (2014)
5. Baarslag, T., Hindriks, K., Jonker, C.: A tit for tat negotiation strategy for real-time bilateral negotiations. In: Ito, T., Zhang, M., Robu, V., Matsuo, T. (eds.) Complex Automated Negotiations: Theories. Models, and Software Competitions, volume 435 of Studies in Computational Intelligence, pp. 229–233. Springer, Berlin (2013)
6. Baarslag, T., Hindriks, K., Jonker, C.: Effective acceptance conditions in real-time automated negotiation. Decis. Support Syst. **60**, 68–77 (2014)
7. Baarslag, T., Hindriks, K.V.: Accepting optimally in automated negotiation with incomplete information. In: Proceedings of the 2013 International Conference on Autonomous Agents and Multi-agent Systems, AAMAS '13, pp. 715–722, Richland, SC (2013). International Foundation for Autonomous Agents and Multiagent Systems
8. Carnevale, P.J.D., Lawler, E.J.: Time pressure and the development of integrative agreements in bilateral negotiations. J. Confl. Resolut. **30**(4), 636–659 (1986)
9. Chen, S., Weiss, G.: OMAC: a discrete wavelet transformation based negotiation agent. In: Marsa-Maestre, I., Lopez-Carmona, M.A., Ito, T., Zhang, M., Bai, Q., Fujita, K. (eds.) Novel Insights in Agent-based Complex Automated Negotiation. Studies in Computational Intelligence, vol. 535, pp. 187–196. Springer, Japan (2014)
10. Faratin, P., Sierra, C., Jennings, N.R.: Negotiation decision functions for autonomous agents. Robot. Auton. Syst. **24**(3–4), 159–182 (1998). Multi-Agent Rationality
11. Fatima, S.S., Wooldridge, M., Jennings, N.R.: Optimal negotiation strategies for agents with incomplete information. In: Revised Papers from the 8th International Workshop on Intelligent Agents VIII, ATAL'01, pp. 377–392. Springer, London (2002)
12. Fatima, S., Wooldridge, M., Jennings, N.R.: Optimal negotiation of multiple issues in incomplete information settings. In: Proceedings of the Third International Joint Conference on Autonomous Agents and Multiagent Systems-Volume 3, AAMAS'04, pp. 1080–1087. IEEE Computer Society, Washington, DC, USA (2004)
13. Fatima, S.S., Wooldridge, M., Jennings, N.R.: Bargaining with incomplete information. Ann. Math. Artif. Intell. **44**(3), 207–232 (2005)
14. Gode, D.K., Sunder, S.: Allocative efficiency in markets with zero intelligence (ZI) traders: market as a partial substitute for individual rationality. J. Polit. Econ. **101**(1), 119–137 (1993)
15. Hao, J., Leung, H.-F.: ABiNeS: an adaptive bilateral negotiating strategy over multiple items. In: Proceedings of the The 2012 IEEE/WIC/ACM International Joint Conferences on Web Intelligence and Intelligent Agent Technology-Volume 02, WI-IAT'12, pp. 95–102. IEEE Computer Society, Washington, DC, USA (2012)
16. Kawaguchi, S., Fujita, K., Ito, T.: Compromising strategy based on estimated maximum utility for automated negotiating agents. In: Ito, T., Zhang, M., Robu, V., Fatima, S., Matsuo, T. (eds.) New Trends in Agent-based Complex Automated Negotiations, Series of Studies in Computational Intelligence, pp. 137–144. Springer, Berlin (2012)

17. Kersten, G.E., Vahidov, R., Gimon, D.: Concession-making in multi-attribute auctions and multi-bilateral negotiations: theory and experiments. Electr. Comm. Res. Appl. **12**(3), 166–180 (2013). Negotiation and E-Commerce
18. Kolomvatsos, K., Anagnostopoulos, C., Hadjiefthymiades, S.: Determining the optimal stopping time for automated negotiations. IEEE Trans. Syst. Man Cybern.: Syst. **99**, 1–1 (2013)
19. Leonardz, B.: To Stop or Not to Stop. Some Elementary Optimal Stopping Problems with Economic Interpretations. Almqvist & Wiksell, Stockholm (1973)
20. Li, C., Giampapa, J., Sycara, K.: Bilateral negotiation decisions with uncertain dynamic outside options. IEEE Trans. Syst. Man Cybern. Part C: Appl. Rev. **36**(1), 31–44 (2006)
21. Lin, R., Kraus, S., Baarslag, T., Tykhonov, D., Hindriks, K., Jonker, C.M.: Genius: an integrated environment for supporting the design of generic automated negotiators. Comput. Intell. **30**(1), 48–70 (2014)
22. Niemann, C., Lang, F.: Assess your opponent: a bayesian process for preference observation in multi-attribute negotiations. In: Ito, T., Zhang, M., Robu, V., Fatima, S., Matsuo, T. (eds.) Advances in Agent-Based Complex Automated Negotiations. Studies in Computational Intelligence, vol. 233, pp. 119–137. Springer, Berlin (2009)
23. Raiffa, H.: The Art and Science of Negotiation. Belknap Press of Harvard University Press (1982)
24. Rosenschein, J.S., Zlotkin, G.: Rules of Encounter: Designing Conventions for Automated Negotiation Among Computers. MIT Press, Cambridge, MA, USA (1994)
25. Rubin, J.Z., Brown, B.R.: The Social Psychology of Bargaining and Negotiation. Academic Press, New York (1975)
26. Rubinstein, A.: Perfect equilibrium in a bargaining model. Econometrica **50**(1), 97–109 (1982)
27. Sanchez-Pages, S.: The use of conflict as a bargaining tool against unsophisticated opponents. ESE Discussion Papers 99, Edinburgh School of Economics, University of Edinburgh (2004)
28. Slembeck, T.: Reputations and fairness in bargaining - experimental evidence from a repeated ultimatum game with fixed opponents. Exp. EconWPA (1999)
29. Stahl, I.: Bargaining Theory. Economic Research Institute, Stockholm (1972)
30. van Krimpen, T., Looije, D., Hajizadeh, S.: Hardheaded. In: Ito, T., Zhang, M., Robu, V., Matsuo, T. (eds.) Complex Automated Negotiations: Theories. Models, and Software Competitions, volume 435 of Studies in Computational Intelligence, pp. 223–227. Springer, Berlin (2013)
31. Williams, C.R., Robu, V., Gerding, E.H., Jennings, N.R.: Iamhaggler: a negotiation agent for complex environments. In: Ito, T., Zhang, M., Robu, V., Fatima, S., Matsuo, T. (eds.) New Trends in Agent-based Complex Automated Negotiations, Series of Studies in Computational Intelligence, pp. 151–158. Springer, Berlin (2012)
32. Williams, C.R., Robu, V., Gerding, E.H., Jennings, N.R.: An overview of the results and insights from the third automated negotiating agents competition (ANAC 2012). In: Marsa-Maestre, I., Lopez-Carmona, M.A., Ito, T., Zhang, M., Bai, Q., Fujita, K. (eds.) Novel Insights in Agent-based Complex Automated Negotiation. Studies in Computational Intelligence, vol. 535, pp. 151–162. Springer, Japan (2014)
33. Wu, M., Mathijs, W., Han, P.: Acceptance strategies for maximizing agent profits in online scheduling. In: Esther, D., Valentin, R., Onn, S., Sebastian, S., Andreas, S. (eds.) Agent-Mediated Electronic Commerce. Designing Trading Strategies and Mechanisms for Electronic Markets, volume 119 of Lecture Notes in Business Information Processing, pp. 115–128. Springer, Berlin (2013)

Handling Agents' Incomplete Information in a Coalition Formation Model

Souhila Arib, Samir Aknine and Thomas Genin

Abstract Coalition formation is a problem of great interest in AI, allowing groups of autonomous rational agents to form suitable teams. Our work specially focuses on agents which are self-interested and want to negotiate for executing actions in their plans. Depending on its capabilities, an agent may not be able to perform actions alone. Then the agent needs to find partners, interested in the same actions, and agree to put their resources in common, in order to perform these actions all together. We propose in this paper a coalition formation mechanism based on: (1) an action selection algorithm, which allows an agent to select the actions to propose and deal with the incomplete information about other agents in the system and (2) a coalition evaluation algorithm, which allows an agent to select a group of agents to perform with these actions. Our coalition evaluation algorithm is designed for structured-preference context, based on the use of the information gathered in the previous interactions with other agents. It allows the agents to select partners, which are more likely interested in the actions. These algorithms are detailed and exemplified. We have studied the quality of the solution, we have implemented and tested them, and we provide the results of their evaluation.

1 Introduction

This paper addresses the coalition formation problem in multi-agent systems, where agents are usually guided by specific objectives they intend to reach. They behave autonomously and are able to communicate directly with each other. This paper focuses mainly on self-interested agents where their objective is to form coalitions

S. Arib (✉)
Université Paris Dauphine, Paris, France
e-mail: Souhila.arib@yahoo.fr

S. Aknine
Université Claude Bernard Lyon 1, Villeurbanne, France

T. Genin
Hedera Technology, Montreuil, France

© Springer International Publishing Switzerland 2016
N. Fukuta et al. (eds.), *Recent Advances in Agent-based Complex
Automated Negotiation*, Studies in Computational Intelligence 638,
DOI 10.1007/978-3-319-30307-9_4

with other agents since they cannot reach their objectives individually. By putting their resources together, a group of agents will be able to perform a set of tasks which no agent of the group can perform by itself.

Coalition formation is widely studied in game theory and economics [1]. This research deals with the different following issues: the search for an optimal coalition structure; find the solution of a joint problem, the agents have to agree on a coalitional action to perform; and division of the value of the generated solution among the coalition members. These activities interact with each other, and the agents should reach agreement on all those issues through negotiations [2].

When an agent forms a coalition its level of satisfaction depends on other members of the coalition. The satisfaction of an agent is represented by ordinal or cardinal preferences over the set of coalitions it can form. Since agents are self-interested, their objective is to form the coalition which has the highest possible rank in their own preference order (in case of cardinal preferences, the coalition which has the maximal possible utility) [3, 4].

In our context, agents plans are used to coordinate their actions and form suitable coalitions. In most coalition formation methods, when negotiating their coalitions the agents focus mainly on the immediate tasks to be executed in order to decide which coalitions to form. Agents relegate the negotiations of the coalitions for their subsequent tasks to later stages of the coordination process.

In the context of cooperative agents, Sandholm et al. [5] developed an anytime algorithm for finding the optimal coalition structure, establishing a worst case bound on the quality of the solution. Rahwan et al. [6] proposed another algorithm for distributing the coalition value calculation among agents.

Most models of coalition formation assume that the values of potential coalitions are known with certainty, implying that agents possess knowledge of the capabilities of their potential partners, or at least that this knowledge can be reached via communication [7]. However the objectives of a cooperative agent are different from the objectives of a self-interested agent.

Other works on coalition formation deal with self-interested agents. In this context, coalition formation is a coordination mechanism, which allows self-interested agents to form groups of interest to jointly achieve complex tasks. Kraus et al. [8] focus on the request for proposal domain in which self-interested agents form coalitions to perform complex tasks in order to obtain a payoff. They propose a protocol as well as proposal strategies. The protocol is used by Shehory [9]. However, the context of these works is far from our context and their strategies cannot be adapted. Although these models have addressed important issues, however they do not deal with the context where agents plans their activities.

This paper tackles this problem and proposes a coalition formation model based on the plans of the agents. An agent plan is any sequence of actions ordered to achieve its current goals.

Brafman et al. [10] introduce a planning games which extend STRIPS-like models of single-agent planning to systems of multiple self-interested agents. The key point is that each agent in planning games can in principle influence the utility of the other agents resulting in global inter-agent dependency within the system. They define classes of strategic games and solution concepts that capture the context of multi-agent planning. They developed two basic models corresponding to variants of planning games, coalition-planning games and auction-planning games. In both, each strategy is a course of actions of a single agent, and a joint strategy is a multi-agent plan of the whole system. This model provides a rich class of structured games that capture forms of interactions between the agents.

In the same context, they consider in [11] the model of planning games with transferable utilities (TU). They have connected the idea of planning games and the classical model of TU coalition games. However, they do not deal with the coalition formation mechanism itself. As a solution concept, they focus on the classical concept of the core. They show that both core-existence and core-membership queries in TU planning games can be answered efficiently, and that the notion of stable cost-optimal planning is tractable when cooperative cost optimal planning is tractable. However, they do not use the actions of each agent plan to derive what coalitions will be formed and do not deal with the negotiation of such coalitions.

Our approach uses the plans of the agents to guide the search for the coalitions to be formed. We develop a new mechanism which allows self interested agents to negotiate their coalitions. In this mechanism we assume that the agents have a limit view of the plans of the other agents in the system.

- First, we propose to use an action selection algorithm based on a new concept which is a *degree of belief* of the agents on the actions they can not perform alone and use it to derive the action probability to be done. This first step of the mechanism allows agents to deal with the incomplete information about the other agents and reduce the search time for the coalitions that can be approved quickly.
- Secondly, we propose to take in a consideration the history of the negotiations to derive the intentions of the other agents in the system. We propose a coalition evaluation algorithm which analyzes the coalition proposals already suggested by other agents thus facilitate the negotiations for the coalitions.

This article is structured as follows. Section 2 introduces the coalition formation problem; Sect. 3 presents the solution concepts of our model; Sect. 4 presents the model description; Sect. 5 presents the experimental study of our model and Sect. 6 provides a conclusion.

2 Problem Formulation

We consider a set of $\mathcal{N}$ agents, $\mathcal{N} = \{a_1, a_2, \ldots, a_n\}$ evolving in a multi-agent system. Each agent has a plan to execute, containing a set of actions $\mathcal{A} = \{b_1, b_2, \ldots, b_m\}$. The objective of each agent is to form coalitions to execute the actions of its plan that it can not perform alone. Each agent can combine into bundles the actions which need the coordination and can be negotiated.

The negotiation for coalition formation between agents is performed without any central truthful algorithm or external decision-maker. Each agent makes its own decisions locally during the negotiation phase. Each agent has different preferences regarding the coalitions and does not have to reveal them explicitly; thus it is unaware of the preferences of other agents and agents have to exchange messages about the coalitions they want to form.

A coalition is formed when all participating agents approve it. This common agreement acts like a contract among its members. To summarize, agents follow a common protocol and have private behavior strategies to make decisions on the actions they undertake during this process.

3 Solution Concepts

Let us begin by recalling some important definitions we use in our coalition formation model and mechanism.

3.1 Definitions and Notations

Definition 1 A coalition c is a nonempty subset of $\mathcal{N} \cdot c \subseteq \mathcal{N}$.

We define $\mathcal{C}$ as the set of all coalitions and $\mathcal{C}_i$ the set of coalitions containing agent a_i. $\mathcal{C}_i = \{c \in \mathcal{C} \mid a_i \in c\}$.

When a coalition c is formed, each agent a_i in c gets a certain satisfaction. For an agent a_i, this satisfaction is defined by a utility function $u_i : \mathcal{C}_i \mapsto R$. This function induces a preference order $\succeq_i$ on coalitions of $\mathcal{C}_i$: a_i prefers a coalition c to c' if and only if $u_i(c) \geq u_i(c')$. We denote it by $c \succeq_i c'$. We denote $u_i(c) = u_i(c')$ by $c \sim_i c'$ and $u_i(c) > u_i(c')$ by $c \succ_i c'$.

Definition 2 A solution of the negotiations is a coalition structure CS which is a partition of $\mathcal{N}$, i.e. a set of coalitions $\{c_1, c_2, \ldots, c_k\}$ verifying that:

- $\forall l \in [1, k], c_l \subseteq \mathcal{N}$,
- $\forall (i, j) \in [1, k]^2, i \neq j, c_i \cap c_j = \emptyset$,
- $\bigcup_{l=1}^{k} c_l = \mathcal{N}$.

The set of all coalition structures is denoted $\mathcal{S}$.

3.2 Action Planning

Action planning provides a set of plans that can include actions to be performed by other agents. Those plans that involve actions for several agents require the collaboration among them and thus may trigger a negotiation process.

Each agent a_i performs action b at time t. A plan is a set of such actions.

Note that as some actions may depend on other agents we may be uncertain about them being actually performed.

Since the agents are self-interest and show what they want in their plans, each agent which wants to coordinate with an other must: (1) recognize the plans of this agent, and (2) select the actions for which they will form coalitions.

The evaluation of the confidence of a plan to be executed depends mainly on the application domain and the beliefs of the agent. In our context, the confidence is the probability estimated by an agent that certain agents will be involved in the plan. If an agent considers that it is not possible to perform a plan because the other agents are not going to do what it intends, it will not take it into consideration and will not engage in negotiations over it. The agent will thus look for another option.

The confidence level of a plan is usually in line with the trust on other agents and on the time horizon set by the latest action in the plan. The more agents in the plan depends upon, the lower its probability of success. To increase confidence on a plan to be executed, agents engage in negotiations to form coalitions in order to execute all the actions that the agent can not perform alone. To do so, we propose to calculate a degree of belief which enables the agents to derive their preferences over the actions to be executed.

In this context, we assume that the satisfaction of an agent (the utility) is obtained from the formed coalitions not only from the execution of each action alone or all the actions of the plan.

3.2.1 Action Selection Algorithm

Before starting a negotiation process, the belief of the agent on an action which will be executed can below. So, to start the negotiation for the coalitions each agent must choose the action which has the highest probability to be executed.

To estimate this probability, we propose to calculate a degree of belief for each action which needs coordination of the other agents in the system. Once this degree is calculated, each agent ranks its results in decreasing order to start the negotiations. This eases the process and reduces the negotiation time.

Given an action b of an agent a_i which needs coordination, the estimation of the degree of belief d_b of the agent a_i on its future execution if the other agents agree to perform it, is defined as:

- If the action b does not figure in the plans of the other agents (agents which are interesting in forming a coalition c to execute the action b, the d_b is given by:

$$d_b(a_i) = P(b/\neg\varphi) \times P(\neg\varphi) + \Gamma(a_i, a_{Ag}, \varphi) \times P(\varphi) \tag{1}$$

where: φ is the proposition of a coordination from the other agents a_{Ag} to execute the action b;

$P(\varphi)$ is the probability that there are agents a_{Ag} which want to coordinate with the agent a_i, it's given by:

$$P(\varphi) = (1 - P(\neg\varphi)) \tag{2}$$

$P(\neg\varphi)$ is the probability that no agent of the system considers that it needs to coordinate with a_i at time t; it is given by:

$$P(\neg\varphi) = (\frac{comp_{\mathcal{N}}}{comp_{\mathcal{N}} + 1}) \tag{3}$$

where $comp_{\mathcal{N}}$ is the number of the competitors of the agent a_i that want to form other coalitions.

$P(b/\neg\varphi)$ is the probability that the action b could be executed without any coordination proposition from the other agents. This probability is given by the agent itself;

$\Gamma(a_i, a_{Ag}, \varphi)$ is the trust that has a_i on the other agents a_{Ag} to do what they have agreed on before [12].

Using the evaluation model of the trust proposed by [13] for the decentralised systems, we propose to evaluate the trust of an agent a_i toward another agent a_j as a weighted measure of its interactions with this agent a_j and the informations gathered from the other agents of the system about their interactions with a_j.

$$\Gamma(a_i, a_j, \varphi) = \frac{w_e \times \Gamma_e(a_i, a_j) + w_r \times \Gamma_r(a_i, a_j)}{w_e + w_r} \tag{4}$$

and:

$$\Gamma(a_i, a_{Ag}, \varphi) = \frac{1}{n} \sum_{Ag=1}^{n} \Gamma(a_i, a_{Ag}) \tag{5}$$

- If the action b figures in the plan of one other agent a_j, the d_b is given by:

$$d_b(a_i) = P(b/\neg\varphi) \times P(\neg\varphi) + \Gamma(a_i, a_j, \varphi) \times P(\varphi) \tag{6}$$

where: $P(b/\neg\varphi)$ is the probability that the action b being executed without any coordination proposition from the agent a_j;

$P(\neg\varphi)$ is the probability that the agent a_j has any preference to coordinate with a_i to execute b;

$\Gamma(a_i, a_j, \varphi)$ is the trust that has a_i on the a_j to execute b;

- If the action b figures in the plans of more than one agent, the d_b is given by:

$$d_b(a_i) = Argmax_{d_b \in D_{b_n}}(D_{b_n}, p_n) \tag{7}$$

where D_{b_n} is the set of the agent's degrees of belief on the other agents for which the action b figures in their plans p_n.

This first step of the mechanism allows the negotiation of the preferred actions of each agent.

Example 1 We consider a set of 5 agent in the system $N = \{a_1, a_2, a_3, a_4, a_5\}$, and let us focus on the agent a_1. We assume that a_1 wants to form coalitions to execute these actions in its plan $\{b_1, b_2, b_4\}$. Firstly, it computes the degree of beliefs about the execution of these actions. As an example, there is the results of computation the degree of belief that the action b_1 being executed. From the Eq. 2 it obtains $P(\neg\varphi) = \frac{4}{5}$, from the Eq. 3: $P(\varphi) = \frac{1}{5}$, $\Gamma(a_1, a_2) = \frac{3}{5}$, $P(b/\neg\varphi) = 0$. So from the Eq. 4: $d_{b_1}(a_1) = \frac{3}{25}$. For b_2, $d_{b_2}(a_1) = \frac{2}{25}$ and for $d_{b_4}(a_1) = \frac{4}{5}$.

So, the preferences of the agent a_1 are: $b_4 \succ_1 b_1 \succ_1 b_2$.

4 Model Description

The coalition formation process we propose is based on two steps: coalition search and negotiation. In this process, we assume that the agents evolve in a system where each agent can not perceive all the actions of the plans of the others, so the concept of the degree of belief described previously will be used to derive the preferences of the agents and start the negotiation.

4.1 Negotiation

Since an agent a_i has calculated his preferences using the algorithm described before, it chooses a coalition c in C_i in which it is interested. c is formed by k agents: a_i itself and $k - 1$ other agents of the system, which are not members of an already formed coalition.

Agent a_i, which starts the negotiation, is called the initiator agent and the $k - 1$ other agents a_j are called solicited agents. Firstly agent a_i sends a proposal containing the coalition c to all the $k - 1$ solicited agents.

When an agent a_i receives a proposal of coalition c at a time $t = t_c$, it can either accept it or refuse it. If forming c brings enough satisfaction to a_i i.e. if the utility $u_i(c)$ is high enough, it accepts it, but if $u_i(c)$ is too low, a_i refuses it. We propose to define a threshold $u_i(Ref)$ such as: (1) if $u_i(c) > u_i(Ref)$ then a_i decides to take part in the coalition (acceptance) and (2) if $u_i(c) < u_i(Ref)$ then a_i refuses to take part in the coalition (withdrawal).

Since the agents are rational, we make two assumptions: (1) A rational agent a_i does not accept any proposal of a coalition c whose utility is lower than the utility $u_i(Ref)$. (2) A rational agent a_i, which has already proposed a coalition c to a group of solicited agents at a time t, will accept to join every coalition c' such as $u(c') > u(c)$.

We have chosen to use a simple protocol in order to avoid restraining the behaviors of the agents too much in their negotiations. The protocol consists of three phases: initialization of the negotiation, negotiation itself, and transmission of the solution.

1. Initialization of the negotiation and transfer of the actions and propositions of coordination. The initiator agent a_i informs agents a_j that it is beginning a new negotiation. Each agent a_i asks the other agents a_j to send their actions, and then deduces the set of actions to be performed. a_i computes its preferences about its actions using the action selection algorithm. After that, it selects the coalitions it wants to propose and gathers these coalitions in groups. Agent a_i sends coalition formation proposals to the solicited agents a_j.

2. The negotiation. Each solicited agent a_j answers with an acceptance if it is interested in the coalition or with a withdrawal if it is not. An agent a_j accepts a coalition if the utility it offers is at least equivalent to that of its reference coalition.

Let $\mathcal{P}_i$ be the sorted set of coalitions of C_i proposed by agent a_i during the process of coalition formation, with an order relation $\lhd_i$ such as: $\forall(c_1, c_2) \in \mathcal{P}_i^2$, $c_1 \lhd_i c_2$ means that c_1 is proposed before c_2, by a_i.

Each agent a_i proposes its proposals following these conditions:

- $\forall c \in \mathcal{P}_i$, $c \in C_i$, i.e. a_i belongs to coalition that it proposes.
- $\forall(c_1, c_2) \in \mathcal{P}_i^2$, $c_1 \lhd_i c_2 \Rightarrow c_1 \succeq_i c_2$ and $c_1 \neq c_2$, i.e. the utility of coalitions of $\mathcal{P}_i$ is decreasing following the order $\lhd_i$ and a coalition is only proposed once.
- $\forall c \in \mathcal{P}_i, c \succ_i \{a_i\}$, i.e., each proposed coalition is strictly preferred to the reference coalition.
- All coalitions are proposed, following the preference order of the agent derived from its plans.

  Once the agent a_i has sent its coalition to its preferred agents, it waits for their answers. It receives possibly new coalition proposals that a_i will use to update its coalitions or generate new ones as shown in section B until a solution is reached or the negotiation fails.

3. Transmission of the solution. If all the solicited agents commit to the coalition, agent a_i sends a confirmation of the coalition formation, the coalition is formed and becomes a part of the final coalition structure. If at least one solicited agent does not commit, agent a_i sends a cancellation message to the other members and the negotiation is cancelled.

4.2 Coalition Evaluation Algorithm

We propose in this section a strategy based on the coalitions that an agent previously received and the coalitions this agent previously proposed. This strategy uses the entire negotiation history. During the coalition formation process, each agent a_i receives some proposals from other agents and sends proposals to others. An agent a_i can assume that proposals it receives are well valued by their initiator agent. Moreover, it knows which agents are interested in its own proposals.

Assuming that preferences of other agents are structured (i.e. are not completely random), an agent a_i can use the history of all negotiations as a source of information that can be used in order to improve its future proposals.

An agent a_i evaluates each coalition following its preference order derived from its plans and proposes a coalition for a_j only if it estimates that the coalition is worthwhile. We propose an algorithm allowing agents to estimate if a coalition is interesting to propose or not.

Algorithm 1: $\gamma(c)$ evaluation

1 **repeat**
2     $j \leftarrow 0$;
3   **if** $\gamma(c_j) < \Delta$ **then**
4         $a_i proposes c_i$;
5 **until** a_i *has formed* c;

We summarize the process in Algorithm 1, where c_j is the coalition of rank j in the preference order of a_i, $\gamma(c)$ is an evaluation function in $[-1,1]$ indicating whether it seems interesting for a_i to propose c: the lower the $\gamma(c)$ is, the more c is interesting, and Δ is a threshold predetermined by the agent or its designer, which has an influence on the number of coalitions that will be selected to be sent or not.

Then we describe in the following section an evaluation function $\gamma(c)$, measuring the degree of interest of a coalition c.

Agent a_i stores all the proposals it has received and sent as well as the responses of the solicited agents. The main idea of this strategy is that if c is similar to a coalition c_1 already sent by a_j or similar to a coalition c_2 previously proposed by a_i to a_j, which was accepted by a_j, then since c_1 or c_2 were previously interesting for a_j, c is supposedly interesting for a_j too.

Conversely, if c is similar to a coalition c_3 previously proposed by a_i to a_j, which was refused by a_j, then since c_3 was previously not interesting for a_j, c is supposedly not interesting for a_j either. To compare c to the previous proposals, a_i uses a measure of similarity between these coalitions.

The similarity between two coalitions c_1 and c_2 is defined as:

$$sim(c_1, c_2) = \frac{|(c_1 \cap c_2)|}{|(c_1 \cup c_2)|} \tag{8}$$

where: $|c|$ corresponds to the cardinal of the coalition c.

To define a degree of trade-off Δ we use the Gini measure as a measure of dispersion that compares pairs of elements, in our case coalitions, of a set.

The Gini measure of the coalitions similarity is given by:

$$\Delta(c_1, c_2) = \frac{1}{|\mathcal{N}|(|\mathcal{N} - 1|)} \sum_{(c_1,c_2)\in C_i} sim(c_1, c_2) \qquad (9)$$

For each solicited agent a_j in c, a_i computes: (1) the similarity between c and all the proposals received from a_j and stores the minimum Δ_i^R; (2) the similarity between c and all the proposals sent to a_i, which have been accepted by a_j and stores the minimum Δ_i^A; (3) the similarity between c and all proposals sent to a_j, which have been refused by a_j and stores the minimum Δ_i^{Re}.

Algorithm 2: Coalition evaluation algorithm

---

1 **begin**
2    $\gamma_i \leftarrow 0$;
3   **forall the** $a_j \in c$ **do**
4       $\Delta^R \leftarrow 1$;
5   **forall the** $c' \in C_{i,j}^R$ **do**
6   **if** $\Delta(c, c') < \Delta^R$ **then**
7             $\Delta^R \leftarrow \Delta(c, c')$;
8   **end**
9       $\Delta^A \leftarrow 1$;
10   **forall the** $c' \in C_{i,j}^A$ **do**
11   **if** $\Delta(c, c') < \Delta^A$ **then**
12             $\Delta^A \leftarrow \Delta(c, c')$;
13   **end**
14   **end**
15       $\Delta^{Re} \leftarrow 1$;
16   **forall the** $c' \in C_{i,j}^{Re}$ **do**
17   **if** $\Delta(c, c') < \Delta^{Re}$ **then**
18             $\Delta^{Re} \leftarrow \Delta(c, c')$;
19   **end**
20   **end**
21   **end**
22       $\gamma_i \leftarrow max(\gamma_i, min(\Delta^R, \Delta^A) - \Delta^{Re})$;
23   **end**
24 **end**

Then a_i computes for each solicited agent a_j, the degree of interest: $\gamma_j(c) = min(\Delta_i^R, \Delta_i^A) - \Delta_i^{Re}$.

When $\gamma_j(c) < 0$, c is closer to a previously interesting coalition for a_j than to an uninteresting one, and when $\gamma_j(c) > 0$, it is the opposite. The coalition c can be formed if all the solicited agents are interested. Then to measure the potential of a coalition c, a_i analyses the degree of the agent which has the highest $\gamma(c) = max_{a_j}(\gamma_j(c))$. Algorithm 2 summarizes these results.

Example 2 Example continued. Let us assume now that a_1 wants to form this coalition to do the first action b_4: $c_1^1 = \{a_1, a_2, a_3, a_5\}$ which has been refused by a_3, a_4 and accepted by a_2. $c_1^2 = \{a_1, a_2, a_3, a_4, a_5\}$ which has been refused by a_2, a_4 and been accepted by both agents a_3 and a_5.

We assume that a_1 has already received these coalitions from the agents a_2 and a_3:

$c_2^1 = \{a_1, a_2\}$ from a_2
$c_2^1 = \{a_1, a_2, a_3, a_5\}$ from a_2
$c_3^1 = \{a_1, a_3, a_5\}$ from a_3

Agent a_1 wants to determine whether $c_1^3 = \{a_1, a_2, a_3\}$ is worth to be proposed. For each solicited agents a_2 and a_3, agent a_1 computes $\gamma_j(c_1^3)$ according to the algorithm described before.

For agent a_2, it computes:

$\Delta(c_1^3, c_2^1) = 0.05$, $\Delta_2(c_1^3, c_2^2) = 0.05 \implies \Delta_2^R = min(0.05, 0.05) = 0.05$
$\Delta(c_1^3, c^1) = 0.05$, $\Delta_2^A = 0.05$
$\Delta(c_1^3, c^2) = 0.03$, $\Delta_2^{Re} = 0.03$
$\gamma_2(c_1^3) = min(0.05, 0.05) - 0.03 = 0.02 \implies c_1^3$ seems not interesting for a_2.

For agent a_3, it computes:

$\Delta(c_1^3, c^1) = 0.025, \implies \Delta_3^R = 0.025$
$\Delta(c_1^3, c^1) = 0.03$, $\Delta_3^A = 0.03$
$\Delta(c_1^3, c^1) = 0.03$, $\Delta_3^{Re} = 0.03$
$\gamma_3(c_1^3) = min(0.025, 0.03) - 0.03 = -0.005 \implies c_1^3$ seems interesting for a_3.
a_2 is the most reluctant agent, so $\gamma(c_1^3) = max(\gamma_2(c_1^3), \gamma_3(c_1^3)) = 0.025 > 0$,
a_1 will not propose c_1^3 according to this strategy.

4.3   The Pareto Optimality of the Solution

Inspired by [14] we present in this section the analysis of the solution.

Let $\mathcal{N}$ a set of agents, A a set of actions declared by N and SSS a set of coalition structures in N.

Definition 3 (*Pareto Dominance*) CS' Pareto dominates CS strictly when $\forall a_i \in N$, $CS' \succeq_i CS$ and $\exists a_j \in N \mid CS' \succ_j CS$.

Definition 4 (*Pareto Optimality*) CS is Pareto optimal when no other coalition structure strictly dominates CS, i.e. when: $\neg(\exists CS' \in \mathcal{S}, CS' \neq CS, \forall a_i \in N, CS' \succeq_i CS$ and $\exists a_j \in N \mid CS' \succ_j CS)$.

Proposition 1 *Let* (N) *a group of agents verifying the following properties:*

(1) Coalitions are validated by their initiator before their formation.

(2) All agents have a strict preference order over coalitions to form. This preference is derived from their plans.

(3) All agents use the proposal process described before.

(4) All agents use an acceptance strategy verifying this property: $u_i(c) \geq u_i(Ref)$.

Then all negotiations lead to the formation of a coalition structure CS which is Pareto optimal.

Proof According to the hypothesis 1 the negotiations lead to a coalition structure CS.

We suppose that CS is not Pareto optimal, thus there is another coalition structure CS_1 such as:

$\forall a_i \in N, CS_1 \succeq_i CS$ and $\exists a_j \in N, CS_1 \succ_j CS$.

We split CS in two disjoint subsets: $CS' = CS \cap CS_1$, the subset of common coalitions to CS and CS_1, and $CS'' = CS \backslash CS_1$, the subset of coalitions of CS which are not in CS_1. $CS_1 \neq CS$, therefore $CS'' \neq \emptyset$.

Since the results of the negotiations are validated by their initiator, coalition structures have been finally formed sequentially. Let c be the first coalition to be formed. According to the definition of CS'' : $\forall a_i \in c, c \neq e_i$ (where e_i is the coalition of agent a_i in CS_1) Let a_j be initiator agent of c. According to the Pareto dominance of CS_1 over CS, $\forall a_i \in c, e_i \succeq_i c$. Therefore either $e_i \succ_i c$, or $e_i \sim_i c$. According to hypothesis (2), preferences are strict and since $e_i \neq c$, necessarily $\forall a_i \in c, e_i \succ_i c$.

Agent a_j, initiator of c, has proposed its coalitions following its preference order (hypothesis 3), therefore since $e_j \succ_j c$, a_j has proposed e_j before having proposed c. We have a contradiction: coalition e_j has not been formed whereas when a_j has proposed e_j, all conditions were met for the formation of e_j.

- All the solicited agents of e_j were present in the system (hypothesis 3): CS'' which has been formed. When a_j has proposed e_j, only agents of coalitions of CS' could have left the process.

  As the structures are disjoint then an agent can belong only to a single structure at the same time; therefore one agent of a coalition of CS can not belong to a coalition of the CS'' and therefore can not belong to e_j because e_j belongs to CS''.

- All the agents a_i of e_j verified $u_i(e_j) > u_i(Ref)$, according to the Pareto optimality and strict preferences, $\forall a_i \in e_j, e_j \succ_i CS_i$. □

5 Expermental Study

We have developed a multi-agent system that meets the constraints of the context described in the previous section. Agents are developed to perform sets of actions organized in their plans. These are represented by interfaces that implement a set of classes and a set of methods such as computing the *degree of belief* of each action.

To evaluate the proposed protocol, we analyzed its performances observing several parameters: the number of exchanged messages, the size of these messages (the number of coalitions they contain) and the time of negotiations. Each measure is a mean of 20 different simulations started with the same parameters.

5.1 Experimental Results

We tested the coalition evaluation strategy on agents implementing our action selection algorithm and we varied the threshold Δ_i. Then we compared it to agents implementing a probabilistic strategy, which consists for an agent of proposing coalitions following its preference order, but with a prefixed probability p_i, that means that a proposal will be sent depending on the probability p_i. The number of proposals sent varies with p_i, when p_i is close to 0, very few proposals are selected to be sent, and when p_i is close to 1, almost all proposals are selected to be sent.

This strategy selects randomly the coalitions to be sent but our strategy selects the coalitions containing the agents potentially interested in the formation of the coalition.

In the Fig. 1 we observe the number of proposals sent based on the parameter p_i when agents use the probabilistic strategy and Δ_i when they use the coalition evaluation algorithm. With a value of Δ_i close to 0, agents are selective and send few proposals.

For values of Δ_i greater than 0.1, almost all proposals are sent. We have compared our coalition evaluation strategy to a probabilistic strategy. The number of proposals sent varies with p_i: if p_i is close to 0, few proposals are selected to be sent, but if p_i is close to 0,1, almost all proposals are sent.

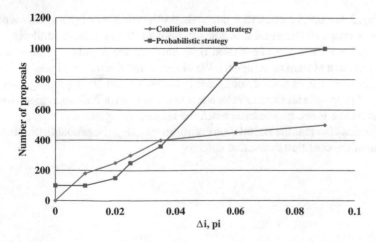

Fig. 1 Number of proposals sent by the agents

S. Arib et al.

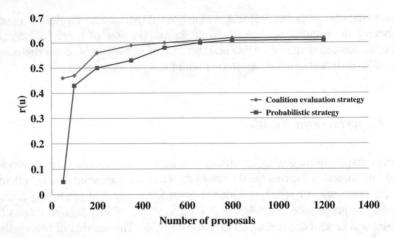

Fig. 2 Ratio of utility

Figure 1 shows the number of proposals sent using the parameters p_i (resp. Δ_i) for agents using the probabilistic strategy and coalition evaluation strategy.

Figure 2 shows the ratio of utility $r(u)$ obtained by the agents at the end of a simulation using the number of proposals sent. In this experimentation we have considered the utility obtained by each agent. We are not interested in the exact value of utility obtained, but we compute the relative value of the utility between the minimal guarantee utility $u(Ref)$ and the maximal possible utility $u(max)$.

We use the ratio $r(u)$, given in Eq. 8. The value of $r(u)$ varies from $r(u(Ref)) = 0$ to $r(u(max)) = 1$.

$$r_i(u) = \frac{u_i(c) - u_i(Ref)}{u_i(max) - u_i(Ref)} \tag{10}$$

In Fig. 2 we do not observe a decrease on the utilities of the agents when the number of proposals decreases. In general, both strategies behave similarly.

In Fig. 3 we show the negotiation time between the agents, we have done the simulations in a system of 50 agents. We observe that the obtained negotiation time in milliseconds with a number of agents between 10 and 50 is growing, due to the number of proposals that agents calculate and exchange for both strategies. However, the search time is acceptable even when the number of agents increases.

We notice also that the negotiations runtime by using the probabilistic strategy is larger than the coalition evaluation strategy.

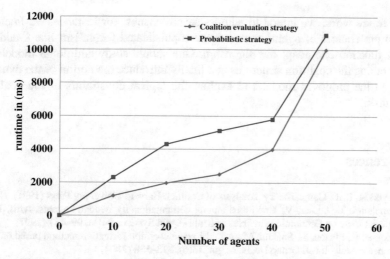

Fig. 3 Negotiation runtime

6 Conclusion

In this paper, we have presented a coalition formation mechanism for self-interested agents. This model uses the plans of the agents and allows them to analyze the coalition proposals already suggested by other agents in order to derive their intentions.

First, we have developed the action selection algorithm based on the *degree of belief* of the agents on actions they can not perform alone. We have introduced the concept of *degree of belief*, in this model, and then we have shown how this degree can be computed based on the different plans of the agents involved in the coalition formation process. The *degree of belief* of an agent on an action allows agents to decide which coalitions should be first formed and deals with the dependencies between the actions of the agents.

Then, we have detailed our model and explained how an agent analyzes the intentions of the other agents during the negotiation to obtain an acceptable solution. We have developed a coalition evaluation algorithm which allows to find the interesting coalitions to propose based on the computation of the similarities between the proposed and already received coalitions.

We have studied the constraints that could be enforced on self-interested agents, in order to form coalitions which guaranty significant solution concepts: Pareto optimality.

Finally we have implemented and tested on a multi-agent system the proposed strategy and have compared it to a probabilistic strategy, then we have analyzed the obtained results. We concluded that the strategy based on the *degree of belief* function and coalition evaluation algorithm allow agents to reduce the number of proposals for the formation of the coalitions while keeping a good utility, which speeds up the process considerably and save time during the coalition formation phase.

In future work, we intend to address several issues, for instance, we intend to model preferences of agents with a more sophisticated formalism like k-additive utility function and adapt our algorithm. One should study additional models for representing the coalition structures and finally introduce the notion of the dynamic plans in the proposed model and explore the logical constraints of the coalition structures.

References

1. Myerson, R.B.: Game Theory: Analysis of Conflict. Harvard University Press (1991)
2. Sandholm, T., Lesser, V.: Coalitions among computationally bounded agents. Artif. Intell. Special Issue on Economic Principles of Multi-Agent Systems **94**(1), 99–137 (1997)
3. Aknine, S., Pinson, S., Shakun, M.: A multi-agent coalition formation method based on preference models. Int. J. Group Decis. Negot. **13**(6), 513–538 (2004)
4. Genin, T., Aknine, S.: Coalition formation strategies for self-interested agents in hedonic games. In: ECAI, pp. 1015–1016 (2010)
5. Sandholm, T., Larson, K., Anderson, M., Shehory, O., Tohm, F.: Coalition structure generation with worst case guarantees. Artif. Intell. **111**(1–2):209–238 (1999)
6. Rahwan, T., Ramchurn, S.D., Dang, V.D., Jennings, N.R.: Near-optimal anytime coalition structure generation. In: IJCAI, pp. 2365–2371 (2007)
7. Shehory, O., Kraus, S.: Methods for task allocation via agent coalition formation. Artif. Intell. **101**(1–2), 165–200 (1998)
8. Kraus, S., Shehory, O., Taase, G.: Coalition formation with uncertain heterogeneous information. In: AAMAS, pp. 1–8. ACM Press (2003)
9. Shehory, O.: Coalition formation towards feasible solutions. Fundam. Informaticae **63**(2–3), 107–124 (2004)
10. Brafman, R., Domshlak, C., Engel, Y., Tennenholtz, M.: Planning games. In: IJCAI (2009)
11. Brafman, R., Domshlak, C., Engel, Y., Tennenholtz, M.: Transferable utility planning games. In: AAAI (2010)
12. Sierra, C., Debenham, J.: Trust and honour in information-based agency. In: AAMAS, pp. 1225–1232. ACM (2006)
13. Huynh, T.D., Jennings, N.R., Shadbolt, N.R.: An integrated trust and reputation model for open multi-agent systems. AAMAS **13**, 119–154 (2006)
14. Bogomolnaia, A., Jackson, M.: The stability of hedonic coalition structures. Games Econ. Behav. **38**(2), 201–230 (2002)

A Multiagent Multilateral Negotiation Protocol for Joint Decision-Making

Romain Caillere, Souhila Arib, Samir Aknine and Chantal Berdier

Abstract We tackle the problem of multilateral negotiation where a group of agents has to make a joint decision. The protocol we propose guides the agents in efficiently coordinating their interrelated negotiations and to coming to an agreement especially when convergence is difficult to obtain in multiagent settings. Negotiation illocutions are formalised to allow interactions between the agents. In addition rules are provided for the agents to manage their interactions. We introduce an illustrative scenario and detail the properties of our protocol. Experimental results show that our protocol guides the agents towards an agreement with reasonable communication and computational complexity.

1 Introduction

Multilateral negotiations are commonly used in situations where interests of several groups, companies, organizations, etc., should converge on a decision and where a set of resources are the common issue. In multilateral negotiation, more than two partners negotiate simultaneously but not in a bilateral way nor via intermediate agents [1, 6]. These negotiations are performed through multiple simultaneous interactions. However in some contexts, although these negotiations involve many participants, negotiations must be guided by those which have either better knowledge of the problem or are more influential and able to provide strategic guidance on significant solutions without necessarily imposing their own choices.

R. Caillere (✉) · S. Aknine
Université de Lyon, Lyon, France
e-mail: romain.cailliere@liris.cnrs.fr

S. Arib
Université Paris Dauphine, Paris, France

C. Berdier
EDU Laboratory Insa of Lyon, Lyon, France

© Springer International Publishing Switzerland 2016
N. Fukuta et al. (eds.), *Recent Advances in Agent-based Complex Automated Negotiation*, Studies in Computational Intelligence 638, DOI 10.1007/978-3-319-30307-9_5

Fig. 1 This figure depicts
the interactions of the agents
in both the inner and external
rings. The *arrows* represent
the direct interactions
between the agents and
dashed lines refer to indirect
interactions (through
overhearing)

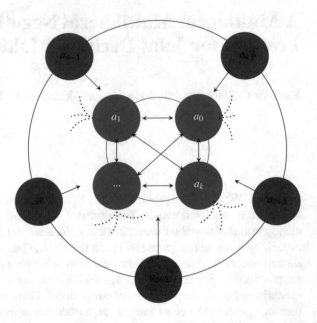

To give an example, the setting up of a tramway line and deciding of its best
route requires a multilateral negotiation. Each participant has his own preferences
and makes his proposals for the line route and the location of stops, however the
most influent participants (regional councilors) involved here have more appropriate
guidance for the decision-makers than others. They can lead this negotiation and
facilitate its convergence owing to their expertise and skills. Thus involving them at
the start of the negotiations in order to early bring their expertise on the problem and
allowing them a wider hearing by the most negotiators before others is an effective
strategy since this may avoid divergences and discussions.

The underlying idea behind our protocol consists in partitioning the agents into
two different sets in order to better control and monitor their interactions during the
negotiation process. We represent these sets of agents by two imbricated rings (cf.
Fig. 1). Each ring has a specific role.

The agents of the inner ring guide the negotiations and those of the external ring
interact with the agents of the inner ring. The purpose of these rings is to allow the
agents to interact in a multilateral way, but also in a controlled manner. Initially, each
agent is assigned to one of these sets, based on various possible assignment strategies.
Each ring gives to its respective agents some allowed behaviors. In the inner ring, the
agents can make their proposals. They can receive the proposals formulated by the
agents in this ring and react by reinforcing, attacking, accepting or refusing them. In
addition, they can receive attack, reinforce and refuse messages from the agents in
both rings.

In the external ring, the agents can overhear the messages exchanged by the agents in the inner ring. Under some conditions, these agents can accept, refuse, reinforce or attack the proposals formulated by the agents in the inner ring. We define some rules for the transitions of the agents from one ring to another. These rules allow all the agents to visit the inner ring where they can defend their own proposals. The negotiation allows each negotiator to suggest possible agreements to the other agents. A proposal can be sent at anytime to the other negotiators if it complies with the submission rules. Inside these rings, there is no predefined order of the interactions for the agents. As we will show, this negotiation framework directs the progression of the interactions and controls the negotiations according to rules accepted by the negotiators. These are the motivations of this protocol.

In the following, Sect. 2 briefly presents the related work. Section 3 presents the application context used to illustrate our mechanism. Section 4 proposes some preliminaries. Section 5 formally describes the protocol. Section 6 details the new protocol's properties. Finally Sects. 7 and 8 present the experimental results and the conclusion.

2   Related Work

Currently some protocols are a generalization of bilateral protocols [8]. However, multilateral negotiations involve additional difficulties, in terms of negotiation runtime, communication and convergence in decision-making that must be dealt with in the design of such protocol. This paper addresses this challenge and proposes a new approach to tackling these negotiations, where each participant interacts with the others by following a set of rules. These rules are defined in order to limit the communications of the agents, computational complexity and to ease the convergence.

The monotonic concession protocol (MCP) [10] proceeds in rounds, where two agents make simultaneous proposals. In the first round, each agent is free to make any proposal. In any subsequent round, agents make a concession and suggest a new deal or refuse to make a concession. An agreement is found when one agent makes a proposal that his opponent rates at least as high as his own current proposal. [4] suggests extending the MCP however he does not provide a specific protocol. [7] proposes a protocol with three roles for the agents: negotiator, morris column and broker. The negotiator agents send proposals, the morris column agent collects these proposals and the broker agent tries to match the collected proposals. However, morris column and broker agents operate as centralized entities.

[5] focuses on studying concession making in negotiations. They propose a theory of concessions and provides its experimental evaluation. [6] proposes a negotiation model applied to an agent-based job marketplace, where they consider fuzzy preferences. Other works addresses argumentation issues in multilateral negotiation.

[1] deal with persuasion where the objective is to arbitrate among conflicting viewpoints. They propose a protocol for multiparty argumentation in which several agents use argumentation systems, and study what outcomes can be reached. These works [2, 3, 9] mainly deal with argumentation aspects through persuasion, merging

arguments, etc. In this paper, we are not concerned with work on argumentation frameworks but with guiding the interactions of the agents through a new negotiation protocol.

3 Case Study

To illustrate the proposed multilateral negotiation protocol, we suggest the case of a negotiation between several participants who have to decide on the positioning of a tramway line and its stops. The participants involved are: regional councilors, representatives of the tramway company, the city mayor, etc. During the negotiation, these participants negotiate the location of the line, the positions of the stops, etc. Each participant has his own preferences concerning the location of the line and the positioning of the stops. Based on this, each participant builds his own proposals.

At the end of the negotiation, the participants have to converge on an agreement where a majority of them and in the best case, all of them, agree on one proposal that would be the solution. The aim of each agent is to ensure that the final solution is the best for him. To come to an agreement, the negotiation must be managed by accurate rules to better guide interactions and arrive at the relevant proposals on which an agreement can be reached. Since each participant intends to defend his interests and opinion, and thus maximize his utility, he will attempt to turn all his suited proposals into solutions. Those rules are necessary to ensure that the negotiations converge on an agreement and this is even more true when there is a large set of participants. In addition, participants in our context can use arguments to convince those who make the decisions, to defend their proposals and attack or reinforce the proposals of other participants.

4 Preliminaries

In the following, let $\mathscr{A}$ be the set of agents involved in the negotiation and $\mathscr{A}_{inn} \subset \mathscr{A}$ (resp. $\mathscr{A}_{ext} \subset \mathscr{A}$) the agents assigned to the inner ring (resp. the external ring), $\mathscr{A}_{inn} \cap \mathscr{A}_{ext} = \emptyset$, $\mathscr{A}_{inn} \cup \mathscr{A}_{ext} = \mathscr{A}$, $|\mathscr{A}| = N$. Each agent has a set of ordered proposals $P_i \in \mathscr{P}$, $\mathscr{P}$ is finite.

For the example, the proposals are the potential locations of the stops. The arguments of the proposals can be either positive to support the proposal, or negative and could then be used by others to weaken the proposal. The details of the syntax and the semantics of these arguments go beyond the scope of this paper.

Let $\mathscr{F}$ be an evaluation function, $\mathscr{F}$ allows the participants to sort their proposals $\mathscr{F} : \mathscr{P} \to \mathbb{R}$. $p_i \succ_a p_j$ means that a prefers the proposal p_i to p_j. $p_i \sim_a p_j$ means that a is indifferent to both p_i and p_j.

Agents can undertake actions in the period of time allowed for the negotiation $\mathscr{T} = [t_0, t_{dl}]$ where t_{dl} is the deadline by which a negotiation will end. To interact, the agents use the following illocutions, where $p_i^k \in P_i$, is the kth proposal of the agent a_i.

- **Propose**($a_i, \mathscr{A}_{inn}, p_i^k, Arg_i^k$): Here $a_i \in \mathscr{A}_{inn}$ sents p_i^k supported by the arguments Arg_i^k to $\forall a_j \in \mathscr{A}_{inn}, i \neq j$.
- **Accept**($a_i, \mathscr{A}_{inn}, p_r^k$): Here $a_i \in \mathscr{A}$ sends an accept for the kth proposal of the agent a_r to $\forall a_j \in \mathscr{A}_{inn}, i \neq r, i \neq j$.
- **Refuse**($a_i, \mathscr{A}_{inn}, p_r^k$): Here $a_i \in \mathscr{A}$ sends a refuse for the kth proposal of the agent a_r to $\forall a_j \in \mathscr{A}_{inn}, i \neq r, i \neq j$.
- **Attack-Proposal**($a_i, \mathscr{A}_{inn}, p_r^k, Arg_r^k, Att_i$) with Att_i, the arguments of a_i attacking the proposal of a_r. Here $a_i \in \mathscr{A}$ sends an attack for p_r^k, supported by the arguments Arg_r^k to $\forall a_j \in \mathscr{A}_{inn}, i \neq r, i \neq j$.
- **Reinforce-Proposal**($a_i, \mathscr{A}_{inn}, p_r^k, Arg_r^k, Rei_i$) with Rei_i, the arguments of a_i reinforcing the proposal of a_r. Here $a_i \in \mathscr{A}$ sends a reinforce for p_r^k to $\forall a_j \in \mathscr{A}_{inn}$, $i \neq j$.

To clarify these illocutions, let us consider the mayor who proposes to place the stop near a school, since this will be practical for the children. To do this, the mayor uses the illocution *Propose*.

Firstly, the representative of the tramway company may not appreciate this suggestion. If he considers that it is not useful to place a stop at this point since it would be costly to secure this site, he answers with the illocution *Attack-Proposal*. Although the representative does not know the price of securing the site, he thinks that the proposal is not relevant. If he has no arguments to attack this proposal he uses the illocution *Refuse*. On the contrary, the regional councilor may consider it as a serious proposal. If the regional councilor has arguments to reinforce the proposal of the mayor, he will answer with the illocution *Reinforce-Proposal*. If he has no arguments, he answers with the illocution *Accept*.

For each illocution, the agents have a set of possible answers. For example, the possible answers of the illocution *Propose* are {*Refuse, Accept, Attack-Proposal and Reinforce-Proposal*}, and the possible answers of *Accept* are {∅}.

Let $\mathscr{M}$ be a triplet $\langle a_i, s \subseteq \mathscr{A}, illocution \rangle$ with $a_i \in \mathscr{A}$, $\mathscr{M}$ represents a move in the negotiation. For example, $m_i = \langle a_1, \mathscr{A}_{inn}, Propose \rangle$ means that for move i, the agent a_1 submits a proposal to all agents in $\mathscr{A}_{inn}$.

Let $\mathscr{L}$ be an ordered set of moves $\langle m_0, m_1, \ldots, m_n \rangle$, that represents all the negotiation process. The illocution used for the first move m_0 must be *Propose*.

Let $\mathscr{R}$ be a set of rules which control the moves of the agents from the inner to the external ring. For example, an agent is allowed to submit a maximum number of Proposal and a maximum number of Refuse.

Let *Result* be an application of $\mathscr{L}$ in {SUCCESS, FAILURE}. It is the result of the negotiation. $Result(\mathscr{L})$ = SUCCESS if the agents end up on an agreement and $Result(\mathscr{L})$ = FAILURE if they do not.

5 The Multilateral Protocol

5.1 Protocol Principle

The negotiation relies on a framework of two rings; in the inner ring, the agents can negotiate and make proposals, while in the external ring the agents can overhear the conversations that have occurred in the inner ring. They can put forward their opinions by making attacks, reinforcing or refusing submitted proposals.

These agents gradually move to the inner ring where they can make their proposals. Negotiations occurs mainly in the inner ring. The agents of the external ring will gradually join the inner ring, as those of the inner ring leave it. If we consider N agents, there will be:

- k agents assigned to the inner ring,
- $N - k$ agents assigned to the external ring

The protocol we propose consists of three phases:

(1) the joining phase which is the phase where the agents join their places in the rings
(2) the negotiation phase in which the agents interact and make their proposals in order to find an agreement
(3) the corrective phase which allows an agent who has not sent all his proposals in the negotiation phase to catch up and submit them under some conditions. This phase occurs only if no agreement is found in the negotiation phase. It takes place over a short period of time so that the agents have only the necessary time needed to conclude

5.1.1 The Joining Phase

There are several strategies to initially assign the agents to the rings. An example is a random assignment of the agents to their rings. In this case and for k places in the inner ring and k' places in the external ring, the probability for an agent to be initially assigned to the central ring is $\frac{k}{N}$, similarly, the probability for an agent to be in the external ring is $\frac{k'}{N}$.

A second feasible strategy is to assign them based on their influence or their roles. The more important an agent is, the earlier he should make his proposals. For instance, the department representative has a higher weight than the sales representative since he is the provider of the funding. The weight represents the influence that an agent has on the process. The weight can be represented by a value between 0 and 1.

Let $weight(a_i)$ be the function which returns the weight of an agent a_i. For 5 agents, 2 places in the inner ring, 3 places in the external ring and

- $weight(a_0) = 0.9$
- $weight(a_1) = 0.8$
- $weight(a_2) = 0.7$
- $weight(a_3) = 0.6$
- $weight(a_4) = 0.5$

we will have the following assignment:

- a_0, a_1 will join the inner ring,
- a_2, a_3, a_4 will join the external ring

5.1.2 The Negotiation Phase

The inner ring channels the interactions of the agents. The agents visit this ring in order to submit their proposals by respecting the transition rules.

Let Δ^p (resp. Δ^r) be the threshold which is the maximum number of *Propose* (resp. *Refuse*) that an agent can send when he is in the inner ring, and δ_i^p (resp. δ_i^r) the number of *Propose* (resp. *Refuse*) that agent a_i has sent. For example, an agent can submit three proposals and answer with a *Refuse* at most twice.

When these thresholds are reached, the agent must leave the inner ring and he is automatically replaced by an agent from the external ring. According to the strategies defined later, when an agent leaves the inner ring, the agent in the external ring who has not yet been through the inner ring is elected to take his place. Even when these thresholds are not reached, an agent cannot stay indefinitely in the inner ring. When his allotted time in this ring is over, he is automatically transferred to the external ring.

Obviously an agent cannot make the same proposal to the same agent twice, and an agent who has already been through the inner ring cannot return to this ring. So, an agent of the external ring is able to move to the inner ring if there is another agent from this inner ring who is obliged to leave it.

To decide which agent from the external ring and who have never been through the inner ring will move to this ring, we suggest different strategies.

For instance:

- random selection of an agent
- choose the agent with the highest weight
- let the agents in the external ring decide on the agent who represents them the best

There are two possible outcomes for negotiation, SUCCESS (the agents come to an agreement) or FAILURE. An agreement is found if a majority $\in [0, 1]$ (value determined before the beginning of the negotiation) of agents accepts one of the submitted proposals. Let $\mathscr{I}_{p^k}$ be the preference of all the agents for a proposal p^k. There are different ways to compute $\mathscr{I}_{p^k}$:

(1) $\mathscr{I}_{p^k} = \frac{|\mathscr{A}_{p^k}^{agreed}|}{N}$, the proportion of agents who agreed, with $\mathscr{A}_{p^k}^{agreed}$ the set of agents who agreed on the proposal p^k

(2) $\mathscr{I}_{p^k} = \frac{\sum weight(a_{p_i^k}^{agreed})}{\sum weight(a_i)}$, with $a_{p^k}^{agreed}$ agent who agreed on p^k

If $\mathscr{I}_{p^k} \geq$ majority, then the process ends on a SUCCESS with the solution p^k.

5.1.3   The Corrective Phase

If no agreement is found during the negotiation phase and the agents in the external ring still have proposals to make, they can submit them using the illocution *Final-Propose*(a_i, $\mathscr{A}_{inn}$, p_i^k, Arg_i^k) with: $a_i \in \mathscr{A}_{ext}$ to $\forall a_j \in \mathscr{A}_{inn}$, $i \neq j$ and $p_i^k \in P_i$ the kth proposal of a_i.

In this phase a compromise may be reached since this phase is regarded as a last chance to come to an agreement and seeing that there is also a limited allotted time for it.

5.2   Behavior of the Agents in the Negotiation and Corrective Phases

In this section, we detail the behavior of the agents during the negotiation.

5.2.1   Behavior of an Agent in the Inner Ring

An agent a_i joins the inner ring at the initiation of the negotiation process or after his election when an agent a_j leaves the inner ring. When a_i arrives in the inner ring, he can begin to submit his proposals while complying with the predefined threshold Δ^p. Then, he evaluates all messages received and decides either to respond by refusing, accepting, attacking or reinforcing the received proposals or by taking into account the attack or reinforcement arguments for the proposals.

If a new agent joins the inner ring, a_i has to add him in his list before sending his next proposals. When a_i reaches the thresholds Δ^p of proposals and Δ^r of refusals or his allotted time in the inner ring is over, he has to leave this ring and move to the external ring. If an agreement is found, the negotiation stops on a SUCCESS. If no agreement is found and all the agents have made their proposals, the negotiation ends on a FAILURE. When a process finishes on a FAILURE, each agent should have gone through the inner ring.

5.2.2 Behavior of an Agent in the External Ring During the Negotiation and Corrective Phases

An agent a_i arrives in the external ring either at the initial set-up of the negotiation or because he has left the inner ring after reaching his thresholds. Each time a_i can overhear the messages formulated by the agents in the inner ring, and those addressed by the agents in the external ring to the agents in the inner ring.

Each message received in a move m_k is processed by a_i. He can refuse (if he has not been through the inner ring), accept (if he has already been through the inner ring), attack or reinforce the proposal in the move m_k. If an agent a_j leaves the inner ring, a_i takes part in the election of an agent a_r, using one of the strategies given above. a_r will join the inner ring and take the place of a_j. If a_i is chosen, he goes to the inner ring. If he is not, he adds the agent a_j who has reached the external ring to his list.

The set of rules which control the illocutions is:

- **R1**: For each $a_i \in \mathscr{A}_{inn}$, the illocution $Propose(a_i, \mathscr{A}_{inn}, p_i^k, Arg_i^k)$ can be made at any time t of the negotiation phase, $t \in \mathscr{T}$ by negotiator a_i for $\forall a_j \in \mathscr{A}_{inn}$ if $\delta_i^p \prec \Delta^p$.
- **R2**: For each $a_i \in \mathscr{A}$, the illocution $Accept(a_i, \mathscr{A}_{inn}, p_j^k)$ can be made at any time $t \in \mathscr{T}$ by a_i for $\forall a_j \in \mathscr{A}_{inn}$: (1) if $a_i \in \mathscr{A}_{inn}$ or (2) if $a_i \in \mathscr{A}_{ext}$ and he has already been through the inner ring. Furthermore the illocution Propose($a_j, \mathscr{A}_{inn}, p_j^k, Arg_j^k$) was made at time $t' \in \mathscr{T}, t' \prec t$.
- **R3**: For each $a_i \in \mathscr{A}$, the illocution $Refuse(a_i, \mathscr{A}_{inn}, p_j^k)$ can be made at any time t of the negotiation phase by negotiator a_i if a_i has not been through the inner and $\delta_i^r \prec \Delta^r$. Furthermore the illocution Propose($a_j \mathscr{A}_{inn}, p_j^k, Arg_j^k$) was made at time $t' \in \mathscr{T}, t' \prec t$.
- **R4**: For each $a_i \in \mathscr{A}$, the illocution $Attack\text{-}proposal(a_i, \mathscr{A}_{inn}, p_j^k, Arg_j^k, Att_i)$ can be made at any time t of the negotiation and corrective phase, $t \in \mathscr{T}$ by negotiator a_i. Furthermore the illocution Propose($a_j, \mathscr{A}_{inn}, p_j^k, Arg_j^k$) was made at time $t' \in \mathscr{T}, t' \prec t$.
- **R5**: For each $a_i \in \mathscr{A}$, the illocution $Reinforce\text{-}proposal(a_i, \mathscr{A}_{inn}, p_j^k, Arg_j^k, Rei_i)$ can be made at any time t of the negotiation and corrective phase, $t \in \mathscr{T}$ by negotiator a_i. Furthermore the illocution Propose($a_j, \mathscr{A}_{inn}, p_j^k, Arg_j^k$) was made at time $t' \in \mathscr{T}, t' \prec t$.
- **R6**: For each $a_i \in \mathscr{A}_{ext}$, the illocution $Final\text{-}Propose(a_i, \mathscr{A}_{inn}, p_i^k, Arg_i^k)$ can be made at any time in the corrective phase if $a_i \in \mathscr{A}_{ext}, \delta_i^p \prec \Delta^p$.

In this protocol, in order to avoid to the agents to conclude with non Pareto-optimal solutions, we forbid them to send accept messages before joining the inner ring. In the following, we prove the consistency of R2.

Proposition 1 *Let a_i be an agent of the inner ring and a_j an agent of the external ring and a_j has not still been through the inner ring. For every agent a_i, if a_j is allowed to accept proposals from a_i, then the solution of the negotiation will not be necessarely Pareto optimal.*

Proof (Sketch) Let a_1, a_2 and a_3 be three agents. Let's assume that the agents are allowed to accept proposals while they are in the external ring even they have not yet been through the inner ring. Let us assume that the preferences of these agents are respectively: $p_1 \sim_{a_1} p_2 \sim_{a_1} p_3 \succ_{a_1} p_4$, $p_2 \sim_{a_2} p_3$, $p_1 \succ_{a_3} p_3 \succ_{a_3} p_2$ and that in the initial assignment, a_1 and a_3 are in the inner ring and a_2 is in the external ring. A proposal becomes a solution if two agents in three come to an agreement on it. The process follows these steps:

(1) a_1 sends the proposal p_2 to a_3 with the message in the move m_0: Propose(a_1, $\{a_3\}$, p_2, $Arg_{a_1}^{p_2}$)
(2) a_3 refuses the proposal p_2 of a_1 with the message in the move m_1: Refuse(a_3, $\{a_1\}$, p_2)
(3) a_2 reacts to the interactions taking place in the inner ring and accepts the proposal of a_1, m_2: Accept(a_2, $\{a_3, a_1\}$, p_2)

Both agents (a_1, a_2) out of the three are in agreement on p_2, p_2 becomes the solution. In this case, p_2 is accepted but p_2 is not Pareto-optimal, however, p_3 is Pareto-optimal. Thus R2 is consistent. $\square$

6 Illustrative Scenario

Let's consider five agents:

- A regional councilor a_0 with $weight(a_0) = 0.9$
- A department representative a_1 with $weight(a_1) = 0.8$
- A tramway company representative a_2 with $weight(a_2) = 0.7$
- A tradesmen's representative a_3 with $weight(a_3) = 0.6$
- A district representative a_4 with $weight(a_4) = 0.5$.

Each agent has an initial ordered list of proposals. A proposal is a potential location defined beforehand for the stops. For example, a_0 builds the proposal p_1. The positive arguments of a_0 for p_1 are {Low costs, Large Crowds} and {Weak connections with the network} are its negative arguments that may weaken it. The negotiation pursues the following steps.

6.1 The Joining Phase

The initial assignment of the agents is made according to their weights. a_0 and a_1 join the inner ring and a_2, a_3 and a_4 are placed in the external ring.

The thresholds allowed for the agent to remain in the inner ring are $\Delta^p = 2$ and $\Delta^r = 1$. This means that an agent can only refuse one proposal and put forward at most two ones.

There is no allotted time considered in this scenario. The election strategy used in this illustrative scenario is: selecting the agent in the external ring who has the highest weight and has never been through the inner ring.

6.2 The Negotiation Phase

After joining their rings, the agents search their optimal solution and negotiate on the set of proposals set $\mathscr{P} = \{p_1, p_2, p_3, p_4, p_5, p_6\}$. Let $\{m_0, m_1, ..., m_{10}\}$ be the moves of the negotiation. We provide some commentary on particular moves selected from the negotiation scenario in Table 1.

First, both agents a_0 and a_1 send their best proposals: m_0, m_1. a_0 proposes the proposal p_1, supported by the arguments $Arg_{a_0}^{p_1}$ and a_1 proposes p_3, supported by the arguments $Arg_{a_1}^{p_3}$.

When the agents receive these proposals, they analyze the associated arguments. Their evaluation function measures the impacts of these arguments on the proposals and sort them in their list of proposals. When the agents in the external ring become aware of these proposals, they add them in their lists.

Next, according to their strategies, a_0 accepts the proposal p_3 (m_2) and a_1 refuses the proposal p_1 (m_3). The agents in the external ring did not react. Now, the proposal p_3 has been accepted by two agents (m_2).

The evaluation of p_3 by a_0 according to his function $\mathscr{F}_{a_0}$, gives him a new preference order for his proposals $p_1 \succ_{a_0} p_3 \succ_{a_0} p_2$. Even if he has accepted p_3, a_0 always prefers p_1. Then, a_1 decides to make another proposal p_6 (m_4) which is refused by a_0 (m_5).

Given that a_1 made one *Refuse* and two *Propose*, he reached the thresholds Δ^p and Δ^r, he thus has to leave the ring and give up his place to one of the three agents of the external ring. The next agent to enter the ring and take the place of a_1 should be selected. a_2 is chosen since he has the highest weight in the external ring.

In the next step, a_2 sends his best proposal (m_6). a_0 and a_2 attack proposals (m_7 and m_8). a_0 attacks with $Att_{a_0} = \{$High maintenance costs$\}$ and a_2 attacks with $Att_{a_2} = \{$Weak connections with the network$\}$.

The evaluation function of a_0 (resp. a_2) lowers p_1 (resp. p_4) in his preference order. His new preference order is $p_3 \succ_{a_0} p_1 \succ_{a_0} p_2 \succ_{a_0} p_4 \succ_{a_0} p_6$ (resp. $p_5 \succ_{a_2} p_4 \succ_{a_2} p_3 \succ_{a_2} p_1 \succ_{a_2} p_6$).

From the external ring a_1 reinforces the proposal p_3 (m_9), considering that he can provide arguments for the proposal he made in m_0. The agent a_2 considers that p_3 is beneficial for him and decides to accept it (m_{10}).

As three agents a_0, a_1, a_2 agreed themselves on setting-up the proposal p_3 and that more than the half of the agents has accepted this proposal, the process ends in a SUCCESS.

Considering that the final ordered list of a_3 (resp. a_4) is $p_4 \succ_{a_3} p_3 \succ_{a_3} p_1 \succ_{a_3} p_6$ (resp. $p_5 \succ_{a_4} p_6 \succ_{a_4} p_3 \succ_{a_4} p_1 \succ_{a_4} p_4$), the negotiations end up with p_3 ranked (1,

Table 1 Example of negotiation moves

m_i	Illocution	$\mathscr{A}_{inn}$	$\mathscr{A}_{ext}$	a_0	a_1	a_2
0	Propose(a_1, {a_0}, p_3, $Arg_{a_1}^{p_3}$)	{a_0, a_1}	{a_2, a_3, a_4}	$p_1 \succ p_2$	$p_3 \succ p_6 \succ p_2$	$p_4 \succ p_5$
1	Propose(a_0, {a_1}, p_1, $Arg_{a_0}^{p_1}$)	{a_0, a_1}	{a_2, a_3, a_4}	$p_1 \succ p_2$	$p_3 \succ p_6 \succ p_2$	$p_4 \succ p_5$
2	Accept(a_0, {a_1}, p_3)	{a_0, a_1}	{a_2, a_3, a_4}	$p_1 \succ p_3 \succ p_2$	$p_3 \succ p_6 \succ p_2 \succ p_1$	$p_4 \succ p_5 \succ p_3 \succ p_1$
3	Refuse(a_1, {a_0}, p_1)	{a_0, a_1}	{a_2, a_3, a_4}	$p_1 \succ p_3 \succ p_2$	$p_3 \succ p_6 \succ p_2 \succ p_1$	$p_4 \succ p_5 \succ p_3 \succ p_1$
4	Propose(a_1, {a_0}, p_6, $Arg_{a_1}^{p_6}$)	{a_0, a_1}	{a_2, a_3, a_4}	$p_1 \succ p_3 \succ p_2$	$p_3 \succ p_6 \succ p_2 \succ p_1$	$p_4 \succ p_5 \succ p_3 \succ p_1$
5	Refuse(a_0, {a_1}, p_6)	{a_0, a_1}	{a_2, a_3, a_4}	$p_1 \succ p_3 \succ p_2 \succ p_6$	$p_3 \succ p_6 \succ p_2 \succ p_1$	$p_4 \succ p_5 \succ p_3 \succ p_1 \succ p_6$
6	Propose(a_2, {a_0}, p_4, $Arg_{a_2}^{p_4}$)	{a_0, a_2}	{a_1, a_3, a_4}	$p_1 \succ p_3 \succ p_2 \succ p_6$	$p_3 \succ p_6 \succ p_2 \succ p_1$	$p_4 \succ p_5 \succ p_3 \succ p_1 \succ p_6$
7	Attack-Proposal(a_0, {a_2}, p_4, $Arg_{a_2}^{p_4}$, Att_{a_0})	{a_0, a_2}	{a_1, a_3, a_4}	$p_1 \succ p_3 \succ p_2 \succ p_4 \succ p_6$	$p_3 \succ p_6 \succ p_2 \succ p_1 \succ p_4$	$p_5 \succ p_4 \succ p_3 \succ p_1 \succ p_6$
8	Attack-Proposal(a_2, {a_0}, p_1, $Arg_{a_0}^{p_1}$, Att_{a_2})	{a_0, a_2}	{a_1, a_3, a_4}	$p_1 \succ p_3 \succ p_2 \succ p_4 \succ p_6$	$p_3 \succ p_6 \succ p_2 \succ p_1 \succ p_4$	$p_4 \succ p_5 \succ p_3 \succ p_1 \succ p_6$
9	Reinforce-proposal(a_1, {a_0, a_2}, p_3, $Arg_{a_1}^{p_3}$, Rei_{a_1})	{a_0, a_2}	{a_1, a_3, a_4}	$p_3 \succ p_1 \succ p_2 \succ p_4 \succ p_6$	$p_3 \succ p_6 \succ p_2 \succ p_1 \succ p_4$	$p_5 \succ p_4 \succ p_3 \succ p_1 \succ p_6$
10	Accept(a_2, {a_0}, p_3)	{a_0, a_2}	{a_1, a_3, a_4}	$p_3 \succ p_1 \succ p_2 \succ p_4 \succ p_6$	$p_3 \succ p_6 \succ p_2 \succ p_1 \succ p_4$	$p_3 \succ p_5 \succ p_4 \succ p_1 \succ p_6$

The columns 1 and 2 represent the moves and their illocutions, column 3 (resp. 4) represents agents in $\mathscr{A}_{inn}$ (resp. $\mathscr{A}_{ext}$). The columns 5, 6 and 7 exemplify the ordered lists of proposals of agents a_0, a_1 and a_2

1, 1, 2, 3) by the agents. This means that p_3 is preferred for a_0, a_1 and a_2, the second (resp. third) choice of a_3 (resp. a_4).

In this scenario, since an agreement has been reached, the agents do not go through the corrective phase.

7 Theorical Analysis of the Protocol

In this section, we present some properties of the protocol. We assume that these following assertions are verified for all properties:

- The agents respect the protocol
- The time to prepare and make a proposal is bounded
- The time to process a message is bounded
- The time to answer a message is bounded
- The time to deliver a message is bounded

Proposition 2 *The communication complexity of the protocol is $\mathcal{O}(mn^2)$.*

Proof (Sketch) Considering a negotiation with m agents and $\mathcal{P}$ the set of all the proposals, with $|\mathcal{P}| = n$, since an agent cannot make the same proposal twice to another agent, and when an agent receives the proposal, he can answer or not, but he should comply with the thresholds Δ^p and Δ^r which limits the number of messages that an agent can send, there are at most $(\Delta^p + \Delta^r) \cdot n^2 \cdot m$ allowed messages for the negotiation.

Arguments are not considered because they are obviously limited and secondary. According to the protocol, an agent is not required to adopt them. In addition an agent has interest to use only plausible arguments which are naturally limited because otherwise its arguments will be discarded by other negotiators.   □

Proposition 3 *The protocol is resistant to scalability.*

Proof (Sketch) Even with a large set of agents, the protocol can manage the negotiations. The number of places in both rings is fixed. Any number of agents can be placed in both of them. To increase the number of participants, agents will only need to increase the runtime, given that more time will be needed for each agent to go into the inner ring.

Furthermore, more time will be necessary for a majority to agree on a solution, if this majority comprises a large set of agents.   □

Proposition 4 *The protocol guarantees that only one proposal will become a solution.*

Proof (Sketch) When an agreement is reached on a proposal p_i the negotiation process ends. So that an agreement on p_i can be found, the majority (defined above) must be reached on p_i. The last agent a_k who accepts p_i is the first to knows that the majority has been reached and he informs the other agents that an agreement has been found and so ends the negotiation process.

However before that every agent knows that the negotiation has ended, the majority may be obtained on another proposal p_j if an agent $a_{k'}$ has sent an accept message. It is thus possible for several proposals to be accepted at the same time. To tackle this problem, when a_k accepts p_i and informs the other agents that the negotiation process ends, every agent knows the time stamps on which p_i and p_j reached the majority.

Agents can simply compare these values and select, the proposal which reached the majority first or the proposal which has been widely accepted.   □

Proposition 5 *For each agent a_i, it is in his best interest to submit his proposals before reaching the corrective phase of the protocol.*

Proof (Sketch) Let a_i be an agent who has left the inner ring without reaching Δ^p because he assumed that he still has a chance to make accept his proposals during the corrective phase.

Since at every moment t during the negotiation phase a proposal could be accepted by agents. If this happens, the protocol ends without reaching the corrective phase and a_i will not have any opportunity to submit his proposals.

Thus, the agents has not interests to wait the corrective phase.   □

Proposition 6 *Based on the protocol, the end of the negotiation process is guaranteed.*

Proof (Sketch) It is easy to show that the protocol guarantees the end of the negotiations. The agents initially placed in the inner ring will necessarily leave for the external ring, either because they have reached their thresholds, or because they have used all their allotted time in the inner ring, given that they have a finite set of proposals. If an agreement is found early in the process, the agents do not need to leave the inner ring.

The external agent who enters the inner ring, will necessarily return to the external ring, unless he is the last agent to visit the inner ring. In this case, when he reaches his thresholds or has no further proposals to submit, he stops submitting proposals.

When no agent has further proposals to submit, the negotiation ends. Given that the number of agents is finite and an agent never returns into the inner ring once he has left it, the negotiation process will necessarily end.   □

Proposition 7 *The protocol is no impacted by the agent breakdown.*

Proof (Sketch) If an agent breaks down in the external ring, this will have no impact on the negotiation. The agent will never give his opinion and he will never be elected to join the inner ring.

If the agent never attains the inner ring, it is as he has never been a participant in the negotiation. If he has already been to the inner ring, the agent will have made his proposals, he will just crease to give his opinion on the other participants' proposals of the other participants. This situation cannot lead the system to deadlock.

If an agent in the inner ring breaks down, since the agents have limited time in this ring, he will necessarily be excluded from this ring. □

8 Experimental Evaluation and Analysis

8.1 Experimental Setting

In this section, we first detail the methodology for analyzing the performances of the proposed negotiation protocol. We then proceed to the empirical study of the protocol. To evaluate the performance of the protocol, a simulation testbed was implemented using JAVA and the API JADE.

We generate the agents and randomly determine their parameters (e.g., their weights, their preferences for the proposals). The mechanism manages the assignment of the agents to the ring, message passing and the negotiation process.

We have performed several tests. In these tests, every agent has a list of proposals. A proposal is ranked between 0 and 1. A value higher than 0.6 indicates that the agent is interested in the proposal and under 0.6 he is not. The agent makes only proposals with a value higher or equal to 0.8. When an agent receives a proposal, he checks:

(1) If the value is under 0.4, the agent will answer with an *Attack-Proposal*. The effect on the sender is a decrease in his value for this proposal if the attack is irrefutable. This decreases the agent's preference for this proposal
(2) If the value is between 0.4 and 0.6, the agent answers with a *Refuse*. There is no effect on the value for the sender
(3) If the value is between 0.6 and 0.8 the agent answers with an *Accept*. There is no effect on the value for the sender but this can trigger the end of the negotiation if the majority is reached on the proposal
(4) If the value is higher than 0.8, the agent answers with a *Reinforce-Proposal* and his value may increase based on used arguments.

We fix some of the parameters given above for the following tests (cf. Figs. 2 and 3). Each agent sorts the proposals in his list.

To elect an agent, they select the one who has the highest weight in the external ring (and has never been through the inner ring). The majority is reached if at least 50 % of the agents agree on a proposal. The proportion of agents in the rings are: 40 % in the inner ring and 60 % in the external ring. To the inner ring are assigned the agents who have the highest weight. The remaining agents are in the external ring. The thresholds are $\Delta^p = 20$ and $\Delta^r = 20$.

In each experiment, the number of agents varies between 10 and 20. In each test, if the majority is reached, an agreement is found and the agents end the negotiation.

We measured the number of messages sent by the agents during the negotiation.

Firstly, we compared the number of all messages sent by all the agents during each negotiation. The experimental results in Fig. 2 show that this number evolves

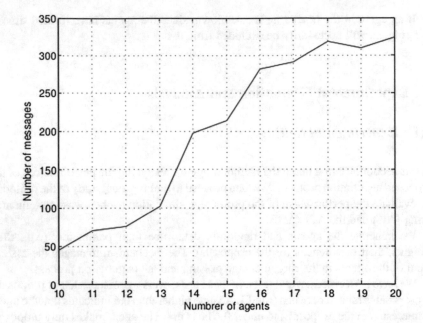

Fig. 2 Number of messages

linearly according to the number of agents involved in the process. This result is intuitive since with the increase in the set of agents, they have more participants to convince, to agree with and reach the majority. Thus agents send and potentially receive more proposals.

8.2 The Quality of the Solution

In the second experiment we observed the performance of the protocol regarding to the quality of the solutions. During the negotiation, the order of the proposals in the list of the agents changes.

To measure the quality of the solution, we compute the average ranking of the proposals which become the solutions. An average ranking at 1 indicates that all the agents agree with the solution.

Formally, the average ranking corresponds to:

$$AR(p) = \frac{1}{N} \sum_{a_i \in \mathscr{A}} rank_{a_i}(p),$$

with $AR(p)$ the function which returns the average rank of the proposal p and $rank_{a_i}(p)$ the function which returns the rank of p in the agent a_i's ordered list.

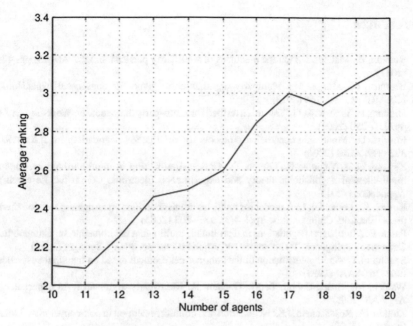

Fig. 3 Average ranking of the final solution

Experimental results in Fig. 3 indicate that the more agents there are in the negotiation the higher the average ranking is. However the proportion of agents in agreement is lower. This observation is intuitive since the more agents there are, the lower the probability that all the agents will agree on a solution is.

9 Conclusion

This paper has presented the design and the implementation of a new multilateral negotiation protocol. The contributions of this paper include the design of a new negotiation framework. The illocutions of the protocol are formally described with their associated rules. An illustrative scenario that provides a detailed example is also given. We have proposed some properties of the protocol. A set of experiments were carried out and the experimental results show that the negotiation protocol we propose guides the agents towards an agreement.

Acknowledgments This work was supported by the LABEX IMU (ANR-10-LABX-0088) of Université de Lyon, within the program "Investissements d'Avenir" (ANR-11-IDEX-0007) operated by the French National Research Agency (ANR).

References

1. Bonzon, E., Maudet, N.: On the outcomes of multiparty persuasion. In: AAMAS, pp. 47–54 (2011)
2. Bromuri, S., Morge, M.: Multiparty argumentation game for consensual expansion. In: ICAART, pp. 160–165 (2013)
3. Dignum, F., Vreeswijk, G.: Towards a testbed for multi-party dialogues. In: Workshop on ACL, pp. 212–230 (2003)
4. Endriss, U.: Monotonic concession protocols for multilateral negotiation. In: AAMAS, pp. 392–399. ACM (2006)
5. Kersten, G.E., Vahidov, R., Gimon, D.: Concession-making in multi-attribute auctions and multi-bilateral negotiations: theory and experiments. Electronic Commerce Research and Applications (2013)
6. Kurbel, K., Loutchko, I.: A model for multi-lateral negotiations on an agent-based job market-place. Electron. Commer. Res. Appl. 4(3), 187–203 (2005)
7. Pałka, P.: Multilateral negotiations in distributed, multi-agent environment. In: Computational Collective Intelligence. Technologies and Applications, pp. 80–89. Springer (2011)
8. Sandholm,T.: An implementation of the contract net protocol based on marginal cost calculations. In: AAAI (1993)
9. Wardeh, M., Bench-Capon, T.J.M., Coenen, F.: Multi-party argument from experience. In: ArgMAS (2009)
10. Zlotkin, G., Rosenschein, J.S.: Negotiation and conflict resolution in non-cooperative domains. In: AAAI (1990)

On the Complexity of Utility Hypergraphs

Rafik Hadfi and Takayuki Ito

Abstract We provide a new representation for nonlinear utility spaces by adopting a modular decomposition of the issues and the constraints. This is based on the intuition that constraint-based utility spaces are nonlinear with respect to issues, but linear with respect to the constraints. The result is a mapping from a utility space into an issue-constraint hypergraph with the underling interdependencies. Exploring the utility space reduces then to a message passing mechanism along the hyperedges by means of utility propagation. The optimal contracts are efficiently found using a variation of the Max-Sum algorithm. We experimentally evaluate the model using parameterized random nonlinear utility spaces, showing that it can handle a large family of complex utility spaces using several exploration strategies. We also evaluate the complexity of the generated utility spaces using the entropy and establish an optimal search strategy allowing a better scaling of the model.

1 Introduction

Realistic negotiation involves multiple and interdependent issues yielding complex and nonlinear utility spaces. Reaching a consensus among a group of agents becomes more difficult as the search space and the complexity of the problem grow.

In this paper, we propose to tackle the complexity of utility spaces used in multi-issue negotiation by rethinking the way they are represented. We think that adopting the adequate representation gives a solid ground to tackle the scaling problem. We address this problem by adopting a representation that allows a modular decomposition of the issues-constraints given the intuition that constraint-based utility spaces are nonlinear with respect to issues, but linear with respect to the constraints. This

R. Hadfi (✉) · T. Ito
Department of Computer Science and Engineering, Graduate School of Engineering,
Nagoya Institute of Technology, Gokiso, Showa-ku, Nagoya 466-8555, Japan
e-mail: rafik@itolab.nitech.ac.jp

T. Ito
e-mail: ito.takayuki@nitech.ac.jp

© Springer International Publishing Switzerland 2016
N. Fukuta et al. (eds.), *Recent Advances in Agent-based Complex
Automated Negotiation*, Studies in Computational Intelligence 638,
DOI 10.1007/978-3-319-30307-9_6

allows us to rigorously map the utility space into an issues-constraints hypergraph with the underling interdependencies. Exploring the utility space reduces then to a message passing mechanism along the hyperedges by means of utility propagation.

Adopting a graphical representation while reasoning about utilities is not new in the multi-issue negotiation literature. In fact, the idea of utility graphs could potentially help decomposing highly nonlinear utility functions into sub-utilities of clusters of inter-related items, as in [1] or [2]. Similarly, [14] used utility graphs for preferences elicitation and negotiation over binary-valued issues. [11] adopts a weighted undirected graph representation of the constraint-based utility space. Particularly, a message passing algorithm is used to find the highest utility bids by finding the set of unconnected nodes which maximizes the sum of the nodes' weight. However, restricting the graph and the message passing process to constraints' nodes does not allow the representation to be descriptive enough to exploit any potential hierarchical structure of the utility space through a quantitative evaluation of the interdependencies between both issues and constraints. In [4], issues' interdependency are captured by means of similar undirected weighted graphs where a node represents an issue. This representation is restricted to binary interdependencies while real negotiation scenarios involve "bundles" of interdependent issues under one or more specific constraints. In our approach, we do not restrict the interdependency to lower-order constraints but we allow p-ary interdependencies to be defined as an hyperedge connecting p issues.

Adopting such graphical representation with its underlying utility propagation mechanism comes from the intuition that negotiation, after all, is a cognitive process involving concepts and associations, performed by supposedly bounded rational agents [9]. And while bearing in mind the fact that cognitive processes perform some form of Bayesian inference [8], we chose to adopt a graphical representation that serves more as an adequate framework for any preference-based space.

The advantage of using this representation is its scalability in the sense that the problem becomes harder for a large number of issues and constraints. But if we can decompose the utility space, we can exploit it more efficiently. Another way to look at this "connectionist" representation is that it can be clustered in ways that can isolate interdependent components, thus, allowing them to be treated separately and even negotiated independently form the rest of the problem.

Another motivation behind the hypergraph representation is that it allows a layered, hierarchical view of any given negotiation scenario. Given such architecture, it is possible to recursively negotiate over the different layers of the problem according to a Top-down approach. Even the idea of issue could be abstracted to include an encapsulation of sub-issues, located in sub-utility spaces and represented by cliques in the hypergraph. Consequently, search processes can help identify optimal contracts for improvement at each level. This combination of separating the system into layers, then using utility propagation to focus attention and search within a constrained region can be very powerful in the bidding process. A similar idea of recursion in the exploration of utility space was introduced by [12] although it is region-oriented and does not adopt a graphical representation of the utility space. We experimentally evaluated our model using parametrized and random nonlinear utility spaces, showing that it can handle large and complex spaces and outperforming previous sampling approaches. The adopted model was also evaluated in terms of the complexity of a family of utility spaces.

Overall, the contribution of the paper could be summarized as following.

- A better representation for nonlinear utility spaces to tackle the complexity problem. It has the merit of being modular and rich, which allows several search strategies to be tested as well as any graph-theoretic analysis.
- An efficient optimization algorithm for optimal contracts search based on message passing. The proposed algorithm outperforms all the other sampling-based methods and provides a better scaling.
- A quantitative assessment of the complexity of nonlinear utility spaces using the *Shannon* entropy [5], and how it could affect the performance of the underlying search algorithm.
- Identification of several search methods and identification of the optimal strategy that minimizes the search cost.

The paper is organized as following. In the next section, we propose the basics of our new nonlinear utility space representation. In Sect. 3, we describe the contracts search mechanisms. In Sect. 4, we provide the complexity study and the optimal strategy. In Sect. 5, we provide some experimental results. In Sect. 6, we conclude and outline the future work.

2 Nonlinear Utility Space Representation

2.1 Problem Formulation

We start from the formulation of nonlinear multi-issue negotiation of [4]. That is, N agents are negotiating over n issues i_k, with $\mathbb{I} = \{i_k\}_{k=1}^{n}$, forming an n-dimensional utility space. The issue k, namely i_k, takes its values from a set $\mathbb{I}_k$ where $\mathbb{I}_k \subset \mathbb{Z}$. A contract $\mathbf{x}$ is a vector of issue values $\mathbf{x} = (x_1, \ldots x_k, \ldots, x_n) \in \mathscr{I}$ with $\mathscr{I} = \times_{k=1}^{n} \mathbb{I}_k$.

An agent's utility function is defined in terms of constraints, making the utility space a constraint-based utility space. That is, a constraint $c_{k \in [1,m]}$ is a region of the total n-dimensional utility space. We say that the constraint c_k has value $w(c_k, \mathbf{x})$ iff it is satisfied by the contract $\mathbf{x}$. That is, when the contract point $\mathbf{x}$ falls within the region of c_k. The utility of an agent for a contract $\mathbf{x}$ is defined as in (1).

$$u(\mathbf{x}) = \sum_{c_{k \in [1,m]}, \, \mathbf{x} \in \sigma(c_k)} w(c_k, \mathbf{x}) \tag{1}$$

$\sigma(c_k)$ is the set of contract points that fall within the region delimited by c_k. In (2), the interval $[a_i, b_i]_{c_k}$ is the set of values taken by issue x_i when it is contained in c_k.

Fig. 1 2-dimensional
nonlinear utility space

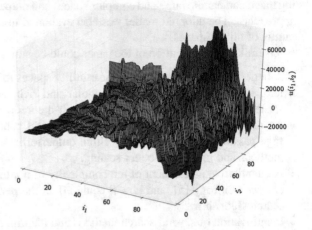

$$\sigma(c_k) = \{ \mathbf{x} \in \mathscr{I} \mid x_i \in [a_i, b_i]_{c_k} \ \forall i \in [1, n] \} \tag{2}$$

Adopting such representation produces a "bumpy" nonlinear utility space with high
points whenever many constraints are satisfied and lower points where few or no constraints are satisfied. For instance, Fig. 1 shows an example of nonlinear utility space
for issues i_1 and i_2 taking values in $\mathbb{I}_1 = \mathbb{I}_2 = [0, 100]$, with $m = 600$ constraints in
$\theta = \{Cube, \ Bell, \ Plane\}$ and where a constraint involves at most 2 issues.

2.2 Alternative Representation

The agent's utility function (1) is nonlinear in the sense that the utility does not have
a linear expression against the contract [4]. This is true to the extent that the linearity
is evaluated with regard to the contract $\mathbf{x}$. However, from the same expression (1) we
can say that the utility is in fact linear, but in terms of the constraints $c_{k \in [1,m]}$. The
utility space is therefore decomposable according to the c_k constraints. This yields a
modular representation of the interactions between the issues and how they locally
relate to each other. In fact, $\sigma(c_k)$ reflects the idea that the underlying contracts are
governed by the bounds defined by c_k once the contracts are projected according to
their issues' components. In this case, the interdependence is not between issues but
between constraints. For instance, two constraints c_1 and c_2 can have in common
one issue i_j taking values respectively from a domain $\mathscr{D}_{c_1}$ if it is in c_1, and values in
$\mathscr{D}_{c_2}$ if it is in c_2, with $\mathscr{D}_{c_2} \neq \mathscr{D}_{c_1}$. Finding the value that maximizes the utility of i_j
while satisfying both constraints becomes harder due to fact that changing the value
of i_j in c_1 changes the instance of i_j in c_2 in a cyclic manner. This gets worse with an
increasing number of issues, their domains' sizes, and the non-monotonicity of the
constraints.

Next, we propose to transform (1) into a modular, graphical representation. Since one constraint can involve one or more multiple issues, we adopt a hypergraph representation.

2.3 Utility Hypergraph

Given $c_{k \in [1,m]}$, we assign to each c_k a factor Φ_k, with $\Phi = \{\Phi_k\}_{k=1}^{m}$. We define the hypergraph G as (3).

$$G = (\mathbb{I}, \Phi) \tag{3}$$

Nodes in $\mathbb{I}$ define the issues and the hyperedges in Φ are the factors (constraints). To each factor Φ_k we assign a neighbors' set $\mathcal{N}(\Phi_k) \subset \mathbb{I}$ containing the issues connected to Φ_k (involved in c_k), with $|\mathcal{N}(\Phi_k)| = \varphi_k$. In case $\varphi_k = 2 \ \forall k \in [1, m]$, the problem collapses to a constraints satisfaction problem in a standard graph. To each factor Φ_k corresponds a φ_k-dimensional matrix defined as $\mathcal{M}_{\Phi_k} = \times_{j=1}^{\varphi_k} [a_j, b_j]$ with $[a_j, b_j] = \mathbb{I}_j$, the domain of issue i_j. This matrix contains all the values that could be taken by the issues in $\mathcal{N}(\Phi_k)$. Each factor Φ_k has a functional form ϕ_k defined as a utility function of the issues in $\mathcal{N}(\Phi_k)$, as in (4).

$$\phi_k : \mathcal{N}(\Phi_k)^{\varphi_k} \to \mathbb{R} \tag{4}$$
$$\phi_k(\mathbf{x}) \mapsto \phi_k(x_1, \ldots, x_j, \ldots, x_{\varphi_k}) = w(c_k, \mathbf{x})$$

As we are dealing with discrete issues, ϕ_k is a linear mapping defined by the matrix $\mathcal{M}_{\Phi_k}$. That is, $\phi_k(x_1, \ldots x_j, \ldots, x_{\varphi_k})$ is simply the $(1, \ldots, j, \ldots, \varphi_k)$th entry in $\mathcal{M}_{\Phi_k}$ corresponding as well to the value $w(c_k, \mathbf{x})$ mentioned in (1).

It is possible to extend the discrete case to the continuous one by allowing continuous issue values and defining ϕ_k as a continuous function. Next, we give few examples about the model usage.

2.3.1 Example 1

Figure 2 illustrates a 2-dimensional utility space with its hypergraph G_2. The issues' domains are $\mathbb{I}_1 = \mathbb{I}_2 = [0, 9]$. G_2 consists of $m = 10$ constraints (red squares) where each constraint involves at most 2 issues (white circles). We note 3 cubic constraints $\{C_j\}_{j=0}^{2}$ and 7 plane constraints $\{P_j\}_{j=0}^{6}$ with parameters $\beta_k, \alpha_k \in [-100, 100]$ and $k \in \{P_0, \ldots, P_6\}$.

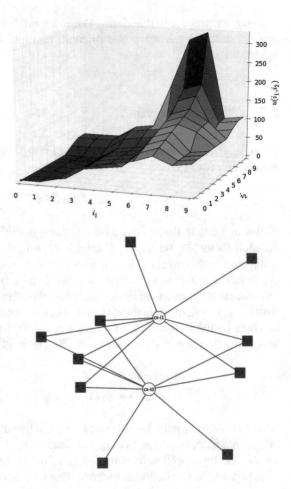

Fig. 2 2-dimensional utility space and its hypergraph

2.3.2 Example 2

Consider the 10-dimensional utility space mapped into the hypergraph G_{10} defined as $G_{10} = (\mathbb{I}, \Phi)$ with $\mathbb{I} = \{i_k\}_{k=1}^9$ and $\Phi = \{\Phi_j\}_{j=1}^7$ as shown in Fig. 3.

Each issue i_k has a set $\mathbb{I}_k = \bigcup_{v \in \mathcal{N}(k)} \mathbb{I}_{k,v}$ where $\mathbb{I}_{k,v}$ is an edge connecting i_k to its neighbor $v \subset \mathcal{N}(k) \in \Phi$. For example, $\mathbb{I}_1 = \bigcup_{v \in \{\Phi_1, \Phi_3, \Phi_6\}} \mathbb{I}_{1,v} = \{[5, 9], [3, 4], [3, 6]\}$.

Constraints are defined by 3 types of geometrical shapes: cubic, plane or bell [10]. For instance, $\Phi_{1,2,3,4}$ are cubic, plane for $\Phi_{5,6}$ and bell for Φ_7. Each constraint is assigned a functional representation used to compute the utility of a contract if it satisfies the constraint by being located in the corresponding hyper-volume. For example, the utility function ϕ_k, defined in (4), corresponds to the functional definition of each constraints, as shown in (5).

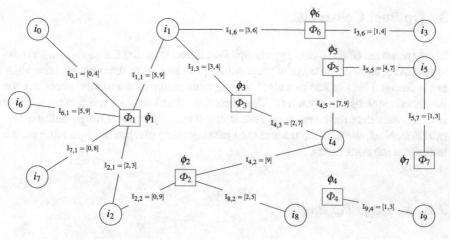

Fig. 3 Issues-constraints hypergraph

$$\phi_k = \begin{cases} \beta_k + \sum_{j=1}^{\varphi_k} \alpha_{k,j} \times v_j(i_j), & \beta_k, \alpha_{k,j} \in \mathbb{Z} \\ v_k \\ V_k \end{cases} \tag{5}$$

A plane constraint Φ_k will be defined using its φ_k-dimensional equation, while a cubic constraint will be assigned the value v_k in case the contract is in the cube. The computation of the utility V_k of a bell shaped constraints is performed as in (6). Herein, δ is the Euclidean distance from the center of the bell constraint to the contract point. Distances are normalized in $[-1, 1]$.

$$V_k = \begin{cases} \beta_j (1 - 2\delta^2) & \text{if } \delta < 0.5 \quad \beta_j \in \mathbb{Z} \\ 2\beta_j (1 - \delta)^2 & \text{if } \delta < 1 \quad \beta_j \in \mathbb{Z} \\ 0 & \text{else} \end{cases} \tag{6}$$

It is important to note that in our usage of the constraints, the agent is required to know the structure of the constraints through their functional definition. However, this is not always the case as the agent might face the situation where the only available assessment tool is a utility function deprived from its internal structure. In this case, we might think of a sampling method allowing us to construct an approximation of the agent's utility space. This could be done in a similar way to what *Markov Chain Monte Carlo (MCMC)* methods or a *Gibbs sampler* could perform in the case of Bayesian problems; especially, with the graphical nature of our representation.

3 Optimal Contracts

The exploration of the utility hypergraph is inspired from the the sum-product message passing algorithm for belief propagation [13]. However, the multiplicative algebra is changed into an additive algebra to support utility accumulation necessary for the assessment of the contracts. The messages circulating in the hypergraph are nothing other than the contracts we are attempting to optimize through utility maximization. Next, we develop the message passing (MP) mechanism operating on the issues and the constraints.

3.1 Message Passing

Next, we consider the issues set $\mathbb{I}$ and a contract point $\mathbf{x} = (x_1, \ldots, x_i, \ldots, x_n) \in \mathscr{I}$. We want to find a contract $\mathbf{x}^*$ that maximizes the utility function defined in (1). Assuming that ϕ_j is the local utility of constraint Φ_j, we distinguish two types of messages: messages sent from issues towards constraints, and messages sent from constraints towards issues. The whole message passing process is an alternation of these two types of messages.

3.1.1 From Issue i_k to Constraint Φ_j

As shown in (7), each message $\mu_{i_k \to \Phi_j}$ coming from i_k to Φ_j is the sum of the constraints' messages to i_k coming from constraints other than Φ_j.

$$\mu_{i_k \to \Phi_j}(i_k) = \sum_{\Phi_{j'} \in \mathcal{N}(i_k) \setminus \Phi_j} \mu_{\Phi_{j'} \to i_k}(i_k) \tag{7}$$

3.1.2 From Constraint Φ_j to Issue i_k

Each constraint message is the sum of the messages coming from issues other than i_k, plus the constraint value $\phi_j(i_1, \ldots, i_k, \ldots, i_n)$, summed over all the possible values of the issues (connected to the constraint Φ) other than the issue i_k.

$$\mu_{\Phi_j \to i_k}(i_k) = \max_{i_1} \ldots \max_{i_{k'} \neq k} \ldots \max_{i_n} \left[\phi_j(i_1, \ldots, i_k, \ldots, i_n) \right.$$
$$\left. + \sum_{i_{k'} \in \mathcal{N}(\Phi_j) \setminus i_k} \mu_{i_{k'} \to \Phi_j}(i_k) \right] \tag{8}$$

The MP mechanism starts from the leaves of the hypergraph, i.e. the issues. At $t = 0$, the content of the initial messages is define according to (9), with $\phi_j'(i_k)$ being the partial evaluation of i_k in the factor Φ_j.

$$\mu_{i_k \to \Phi_j}(i_k) = 0 \quad \text{and} \quad \mu_{\Phi_j \to i_k}(i_k) = \phi_j'(i_k) \tag{9}$$

In a negotiation setting, it is more common that the agent requires a collection, or *bundle*, of the optimal contracts rather than one single optimum. Particularly, in order to engage in a bidding process, the agents needs the optimal subset of the utility space. In order to find such collection, we should ensure a caching mechanism allowing each node in the hypergraph to store the messages that have been sent to it from the other nodes. That is, the cached messages will contain the summed-up utility values of the underlying node's instance. This is performed every time the operation *max* is called in (8) so that we can store the settings of the adjacent utility (and contract) that led to the maximum. Once ordered, such data structure allows us to generate the right representation for the bidding process. In the next section, we algorithmically provide the message passing mechanism.

3.2 Utility Propagation Algorithm

The main algorithm, Algorithm 1, operates on the hypergraph nodes by triggering the message passing process. Despite the fact that we have 2 types of nodes (issues and constraints), it is possible to treat them abstractly using *MsgPass*, in line 6. The resulting bundle is a collection of optimal contracts with utility greater or equal to the agent's reservation value *rv*. The message passing routine, *MsgPass*, is instantiated depending on the types of the source and destination nodes:

From Issue to Constraint. The issue's message to a factor (or constraint) is the element-wise sum of all the incoming messages from other factors.

From Constraint to Issue. The factor's message to a targeted issue is done by recursively enumerating over all variables that the factor references (8), except the targeted issue. This needs to be performed for each value of the target variable in order to compute the message. If all issues are assigned, the values of the factor and of all other incoming messages are determined, so that their sum term is compared to the prior maximum. The resulting messages, stored in *bundle*, contain the values that maximize the factors' local utility functions.

It is possible to avoid the systematic enumeration by adding a local randomization to the issue that the factor is referencing. Additionally, we can exploit the structure of the constraint though its function's monotonicity. That is, by optimizing the constraint locally and providing the optimal sub-contracts as messages. However, optimizing locally does not produce a global optimization due to the interdependence between constraints (example in Sect. 2.2).

Algorithm 1 Main Algorithm

Algorithm: Utility Propagation

Input: $G = (\mathbb{I}, \Phi)$, rv, $mode$, ρ
Output: Optimal contracts (bundle)

1 **begin**
2   **for** $i = 1 \to (\rho \times |\mathbb{I} \cup \Phi|)$ **do**
3   **if** $mode$ is *Synchronous* **then**
4   **foreach** $v_{src} \in \mathbb{I} \cup \Phi$ **do**
5   **foreach** $v_{dest} \in v_{src}.Neighbors()$ **do**
6            $v_{src}.MsgPass(v_{dest})$

7   **else if** $mode$ is *Asynchronous* **then**
8        $v_{src}, v_{dest} \leftarrow rand_2([1, |V|]), v_{dest} \neq v_{src}$
       $v_{src}.MsgPass(v_{dest})$

9    $bundle \leftarrow \emptyset$
   foreach $i \in \mathbb{I}$ **do**
10      $bundle[i] \leftarrow \emptyset$

11      $\iota \leftarrow \cup_{j \in i.instances()}[j.min, j.max]$
     $\mu^* \leftarrow k^* \leftarrow -\infty$
     $\mu \leftarrow i.getmax()$
     foreach $k = 1 \to |\mu|$ **do**
12   **if** $\mu^* < \mu[k]$ **then**
13          $\mu^* \leftarrow \mu[k]$
         $k^* \leftarrow k$
         if $\mu^* \geq rv$ **then**
14            $bundle[i] \leftarrow bundle[i] \cup \iota[k^*]$

15   **return** $bundle$

3.3 Propagation Strategies

The propagation, or circulation of the messages in G could be defined according to a particular strategy with respect to the hypergprah topology. For example, lines 3 in Algorithm 1 refers to a systematic or synchronous way of choosing the nodes. Line 7 corresponds to an asynchronous, or randomized way of selection the sources and the destinations. In the following we will adopt the asynchronous mode of messages transmission.

4 Complexity

Before the evaluation of the hypergraphical representation and the utility propagation algorithm, it is important to identify the criteria that could affect the complexity of the utility space and thus the probability of finding optimal contract(s). To this end, we start by defining the parameters that could have an impact the complexity of the preference spaces. These parameters are also used for the generation of the scenarios.

1. n: number of issues.
2. m: number of constraints, hyperedges, or factors.
3. π: for a constraint Φ_k, $\pi(\Phi_k)$, refers to the number of issues involved in Φ_k, making it unary ($\pi(\Phi_k) = 1$), binary ($\pi(\Phi_k) = 2$), ternary ($\pi(\Phi_k) = 3$), or $\pi(\Phi_k)$-ary in the general case. π is defined as in (10).

$$\pi : \mathscr{C} \to [1, n] \tag{10}$$
$$\Phi_k \mapsto \mathscr{N}(\Phi_k)$$

4. Domain sizes of the issues $\prod_{k=1}^{n} |\mathbb{I}_k|$

A particular utility space (respectively hypergraph) profile is a tuple of the form (n, m, π). This parametrization will be used in the study of the complexity. In a profile, π must meet the consistency condition (11).

$$\pi(\Phi_k) \leq n \leq m \times \pi(\Phi_k), \quad \forall k \in [1, m], \quad \pi \in [1, n]^{\mathscr{C}} \tag{11}$$

Such condition prevents cases like attempting to have an 8-ary constraints in a 5-dimensional utility space. The complexity of a particular profile (n, m, π) is performed using the information theoretical notion of entropy $H(\pi)$ as an application to the general case of cognitive complexity of cognitive graphs [5, 6]. Herein, the information theoretical complexity is meant to capture the eventual duration time of any search algorithm.

For instance, assuming that we have 10 strategies $\pi_{k \in [0,9]}$ where each π_k is either a uniform distribution $\mathscr{U}$, a deterministic distribution $\mathscr{D}$ or a power-law distribution $\mathscr{PL}$. If the search algorithm is taken to be AsynchMP, then the duration time and the complexity of the underlying strategies are illustrated in Fig. 4. The first observation is that both entropy and the duration fluctuate similarly, describing the same topology of the underlying strategy. Secondly, $\mathscr{U}$ is the most complex structure, since it possess the highest entropy/duration as opposed to $\mathscr{D}$ and $\mathscr{PL}$.

It is possible to think about the complexity of strategy $\mathscr{U}$ from two standpoints. An analytical or cartesian view of the problem reduces the complexity to high-dimensionally [3]. However, a connectionist or graphical view sees the problem as a complete graph with one strong component having the highest number of possible connections. Both views reflect the difficulty of the search problem.

4.1 Optimal Strategy

Instead of the asynchronous mode of AsynchMP (randomly picking v_{src} and v_{dest}), we propose to use the distribution π as a prior, allowing us to optimize the message passing algorithm by taking into consideration certian topologies. For example,

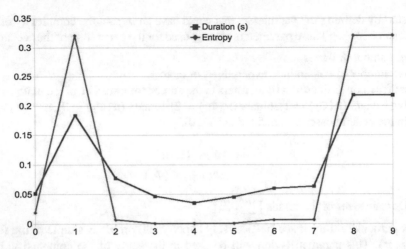

Fig. 4 $H(\pi_k)$ and $\Delta(\pi_k)$ for $\pi_{k \in [0,9]} \in \{\mathscr{U}, \mathscr{D}, \mathscr{PL}\}$

adopting a strategy $\pi \sim PL$ allows us to focus on the hubs of the hypergrpah i.e. factors with large numbers of issues. Let us call the new strategy AsynchMPi, consisting in performing the message passing process within a set $\sigma_1 \subset (\Phi \bigcup \mathscr{I})$ of high degree nodes.

For a specific profile (n, m, π), and for two strategies AsynchMP and AsynchMPi, the idea is to see which one converges to the optimium faster, while being certain that both will find this optimum. That is, the same profile will be traversed and explored with different distributions throughout time.

In the following, we show how the set σ_1 is constructed.

1. Generate the sequence c_j, defined as in (12). The sequence c_j has the property of having the majority of its points clustered in the upper portion of the domain. It will be used to map the set of high degree nodes.

$$
c_j = \begin{cases} \frac{1}{2} & \text{if } j = 0 \\ \\ \frac{1}{2} + \frac{1}{2} \times c_{j-1}^2 & \text{if } j \in [1, (n+m) \times 10] \end{cases} \tag{12}
$$

2. Uniformly sample r points from c_j. Result is $\mathscr{U}(c_j)$
3. Generate the sequence s_π containing $\{\pi(v_i)\}_i$ by decreasing order (13). The sequence s_π approximates a power-law distribution.

$$
s_\pi = \{i \dots n' \mid \pi(v_i) \le \cdots \le \pi(v_{n'})\} \tag{13}
$$

4. Generate the set s_σ according to Algorithm (2).

Algorithm 2 Generation of high degree nodes s_σ

Input: Sequences $\mathscr{U}(c_j)$ and s_π
Output: Set s_σ of high connectivity nodes
begin

    $s_\sigma \leftarrow \emptyset$

    for $c \in \mathscr{U}(c_j)$ **do**

        if $c < \lceil c \rceil - \frac{1}{2}$ **then**

            $s_\sigma \leftarrow s_\sigma \cup \lceil c \rceil - 1$

        else

            $s_\sigma \leftarrow s_\sigma \cup \lceil c \rceil$

    return s_σ

---

5 Discussion

The generation of the hypergraph is performed using Algorithm 3. Depending on the nature of π, a particular topology will be generated. A uniform π generates a complete hypergraph, a power-law π generates a scale-free hypergraph and so on.

In the following we evaluate both SynchMP and AsynchMPi for 5 profiles $(100, 100, \pi_i)$ with $i \in [1, 5]$ and $\pi_i(\Phi_k) \leq p_i$, for $p_i \in [5, 10]$ and $k \in [1, 100]$.

As shown in Fig. 6, both strategies give the same expected optimal utilities. However, Fig. 7 shows that restricting the message passing process to the high degree nodes (Fig. 5) results in a drastic decrease in the duration of the search time. We

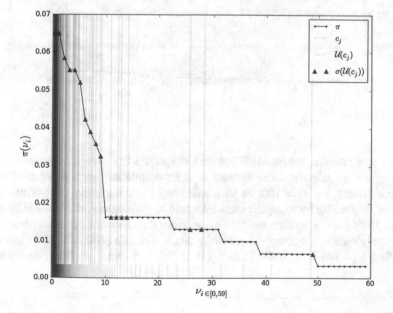

Fig. 5 Generation of s_σ for $n, m, p, r = (10, 50, 5, 17)$

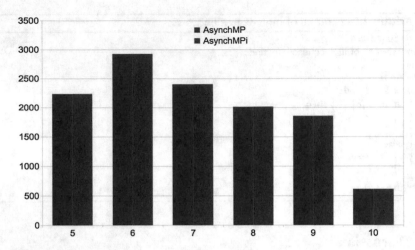

Fig. 6 Utility

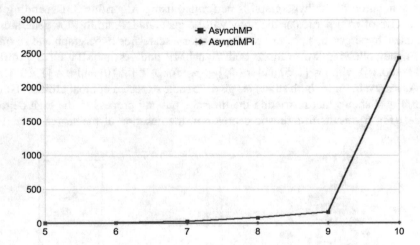

Fig. 7 Duration

additionally observe that for small connectivity values ($\pi_1 = \pi_2$), the search process takes approximately the same amount of time despite the large number of issues and constraints $n = m = 100$. In fact, assessing the complexity of a utility space (respectively utility hypergrpah) must take into consideration the connectivity function π. In this sense, neither the dimension n nor the number of constraints m could objectively reflect this complexity. For instance, a profile (100, 100, 1) is less complex then a profile (10, 10, 2). In fact, (100, 100, 1) is not a nonlinear utility space because $\pi(\Phi_k) = 1 \; \forall k$. That is, each constraint contains one unique issue i.e. the whole utility is reduced to a sum of the partial utilities of the individual issues with, $n = m$. Thus, the whole problem becomes linear with utility function (14),

$$u(i_1, \ldots, i_n) = \sum_{k=1}^{n} \phi_k(i_k) \qquad (14)$$

with ϕ_k being the utility relative to constraint c_k. We note that the previous case is a degenerate case, and that generally, we ought to generate constraints with cardinalities greater or equal to 2.

Algorithm 3 Utility Hypergraph Generation

Algorithm: ParamRandHGen

Input: n, m, π
Output: $G(\mathbb{I}, \Phi)$
begin
  $[\beta_{min}, \beta_{max}] \leftarrow [1, 100]$ // constants
  $[\alpha_{min}, \alpha_{max}] \leftarrow [0, 1]$ // slopes
  $[b_{min}, b_{max}] \leftarrow [0, 9]$ // bounds

  $\Phi \leftarrow [\emptyset] \times m$ // init constraints set
  for $k = 1 \rightarrow m$ **do**
    $\Phi[k].\theta \leftarrow rand(\{cube, plane, bell\})$
    if $\Phi[k].\theta = plane$ **then**
      $\alpha \leftarrow [0] \times n$

      for $j = 1 \rightarrow n$ **do**
        $\alpha[j] \leftarrow rand([\alpha_{min}, \alpha_{max}])$
      $\Phi[k].\alpha \leftarrow \alpha$
    if $\Phi[k].\theta \in \{bell, cube\}$ **then**
      // similar method; refer to (5) or (6)

    $\Phi[k].\beta \leftarrow rand([\beta_{min}, \beta_{max}])$
    $\mu \leftarrow rand([1, n])$
    $\mathbb{I} \leftarrow \emptyset$

    while $|\mathbb{I}| \neq \mu$ **do**
      $\iota \leftarrow \pi(k)$

      if $\iota \notin \mathbb{I}$ **then**
        $\mathbb{I} \leftarrow \mathbb{I} \cup \iota$
    for $j = 1 \rightarrow \mu$ **do**
      $\mathbb{I}[j].a \leftarrow rand([b_{min}, b_{max}])$ $\mathbb{I}[j].b \leftarrow rand([\mathbb{I}[j].a + \epsilon, b_{max}])$
    $\Phi[k].\mathbb{I} \leftarrow \mathbb{I}$
  return Φ

6 Conclusion

We introduced a new representation of utility spaces based on hypergraphs that allows a modular decomposition of the constraints and issues. The exploration and search for optimal contracts is performed based on a message passing mechanism outperforming the sampling based optimizers. Additionally, the model was evaluated in terms of complexity assessment showing that power-law topologies have lower complexity. Consequently, we provided an exploration strategy that searches the hypergrpah based on a power-law topology. Results show that such strategy outperforms drastically the synchronous systematic message passing strategy.

As a future work, we intend to exploit the structure of the hypergraphs by proposing an hierarchical exploration scheme and evaluate it in a hierarchical negotiation scenario. Additionally, we intend to study the interdependence between the issues as to assess their importance and influence on the overall contract.

References

1. Bacchus, F., Grove, A.: Graphical models for preference and utility. In: Proceedings of the Eleventh Conference on Uncertainty in Artificial Intelligence, UAI'95, pp. 3–10. Morgan Kaufmann Publishers Inc., San Francisco, CA, USA (1995)
2. Chajewska, U., Koller, D.: Utilities as random variables: density estimation and structure discovery. In: Proceedings of the Sixteenth Annual Conference on Uncertainty in Artificial Intelligence (UAI-00), pp. 63–71 (2000)
3. Donoho, D.L.: High-dimensional data analysis: the curses and blessings of dimensionality. In: American Mathematical Society Conference. Mathematical Challenges of the 21st Century (2000)
4. Fujita, K., Ito, T., Klein, M.: An approach to scalable multi-issue negotiation: decomposing the contract space based on issue interdependencies. In: Proceedings of the 2010 IEEE/WIC/ACM International Conference on Web Intelligence and Intelligent Agent Technology-Volume 02, WI-IAT'10, pp. 399–406. IEEE Computer Society, Washington, DC, USA (2010)
5. Hadfi, R., Ito, T.: Cognition as a game of complexity. In: Proceedings of 12th International Conference on Cognitive Modeling (ICCM) (2013)
6. Hadfi, R., Ito, T.: Uncertainty of cognitive processes with high-information load. Procedia Soc. Behav. Sci. 97(0), 612–619 (2013). The 9th International Conference on Cognitive Science
7. Ito, T., Hattori, H., Klein, M.: Multi-issue negotiation protocol for agents: exploring nonlinear utility spaces. In: Proceedings of the 20th International Joint Conference on Artificial Intelligence (IJCAI-2007), pp. 1347–1352 (2007)
8. Kwisthout, J., van Rooij, I.: Bridging the gap between theory and practice of approximate bayesian inference. Cogn. Syst. Res. 24(0), 2–8 (2013). Cognitive Systems Research: Special Issue on ICCM2012
9. Lin, R., Kraus, S., Wilkenfeld, J., Barry, J.: Negotiating with bounded rational agents in environments with incomplete information using an automated agent. Artif. Intell. 172(6–7), 823–851 (2008)
10. Lopez-Carmona, M.A., Marsa-Maestre, I., De La Hoz, E., Velasco, J.R.: A region-based multi-issue negotiation protocol for nonmonotonic utility spaces. Comput. Intell. 27(2), 166–217 (2011)
11. Marsa-Maestre, I., Lopez-Carmona, M.A., Velasco, J.R., de la Hoz, E.: Effective bidding and deal identification for negotiations in highly nonlinear scenarios. In: Proceedings of

The 8th International Conference on Autonomous Agents and Multiagent Systems-Volume 2, AAMAS'09, pp. 1057–1064. International Foundation for Autonomous Agents and Multiagent Systems, Richland, SC (2009)

12. Marsa-Maestre, I., Lopez-Carmona, M.A., Carral, J.A., Ibanez, G.: A recursive protocol for negotiating contracts under non-monotonic preference structures. Group Decis. Negot. **22**(1), 1–43 (2013)
13. Pearl, J.: Probabilistic Reasoning in Intelligent Systems: Networks of Plausible Inference. Morgan Kaufmann Publishers Inc., San Francisco, CA, USA (1988)
14. Robu, V., Somefun, D.J.A., La Poutre, J.A.: Modeling complex multi-issue negotiations using utility graphs. In: Proceedings of the 4th International Joint Conference on Autonomous Agents and Multi-Agent Systems (AAMAS 2005), pp. 280–287 (2005)

Negotiations in Holonic Multi-agent Systems

Rahmatollah Beheshti, Roghayeh Barmaki and Nasser Mozayani

Abstract Holonic multi-agent systems (HOMAS) have their own properties that make them distinct from general multi-agent systems (MAS). They are neither like competitive multi-agent systems nor cooperative, and they have features from both of these categories. There are many circumstances that holonic agents need to negotiate. Agents involved in negotiations try to maximize their utility as well as their holon's utility. In addition, holon's Head can overrule the negotiation whenever it wants. These differences make defining a specific negotiation mechanism for holonic multi-agent systems more significant. In this work, holonic systems are introduced at the beginning; and then different aspects of negotiation in these systems are studied. We especially try to introduce the idea of *holonic negotiations*. A specific negotiation mechanism for holonic multi-agent systems is proposed which is consistent with the challenges of HOMAS.

Keywords Multi-agent systems · Negotiation · Holonic

1 Introduction

Negotiation techniques are used to overcome conflicts and coalitions, and to come to an agreement among agents, instead of persuading them to accept a ready solution [13]. In fact, negotiation is the core of many agent interactions, since it is often unavoidable between different project participants with their particular tasks and

R. Beheshti (✉) · R. Barmaki
Department of EECS, University of Central Florida, Orlando, USA
e-mail: beheshti@eecs.ucf.edu

R. Barmaki
e-mail: barmaki@eecs.ucf.edu

N. Mozayani
School of Computer Engineering, Iran University of Science and Technology,
Tehran, Iran
e-mail: mozayani@iust.ac.ir

© Springer International Publishing Switzerland 2016
N. Fukuta et al. (eds.), *Recent Advances in Agent-based Complex
Automated Negotiation*, Studies in Computational Intelligence 638,
DOI 10.1007/978-3-319-30307-9_7

domain knowledge whilst they interact to achieve their individual objective as well as the group goals. The importance of negotiation in MAS is likely to increase due to the growth of fast standardized communication infrastructures, which allow, separately, designed agents to interact in an open and real-time environment and carry out transactions safely [19].

Negotiations in MAS can be divided into two main categories: Negotiations in competitive and cooperative MAS. Competitive MAS refers to systems that agents are fully self-interested, and want to maximize their own pay-offs. In cooperative environments, agents usually care about their pay-off and also others' [17]. Holonic multi-agent systems are not fully competitive or cooperative. They are similar to semi-cooperative MAS [8], but there are some critical differences between these two. Few work has been done in the field of holonic multi-agent systems. In the same manner, negotiations in HOMAS are not studied as a separate phenomenon that much.

In this paper, first of all we provide a brief overview of HOMAS in Sect. 2. Then, in Sect. 3 negotiations in HOMAS are studied and the differences that make them distinct from the rest are illustrated. Section 4 is devoted to proposal of a specific negotiation mechanism for HOMAS, and in Sect. 5 the experimental results are provided. Finally, Sect. 6 includes concluding remarks.

2   Holonic Multi-agent Systems

The theory of holonic structures was proposed by Arthur Koestler in 1967. This theory implies a structure which is composed of components named "Holons". A holon is a self-similar or fractal structure that is stable and coherent and consists of several holons as sub-structures [12]. Each holon may contain several subholons, and might be part of a greater holon itself. In this manner a hierarchical structure forms. The organizational structure of a holonic society or holarchy, offers advantages that the monolithic design of most technical artifacts lacks: They are robust in the face of external and internal disturbances and damage [1, 10]. They are efficient in their use of resources, and they can adapt to environmental changes [11]. The concepts of fractal and holonic system design in manufacturing were proposed to combine top-down hierarchical organizational structure with decentralized control, which takes the bottom-up perspective [18].

Within the multi-agent systems domain, holonic multi-agent systems are a special category which are based on holonic structures introduced above. In these systems, several agents join together and make a holon. Form the external view, each holon is quite similar to a single agent. It has common properties of an agent. So, we can use the terms 'holon agent' and 'agent' interchangeably. Figure 1 shows a simple scheme of HOMAS. Each holon usually has a representative called "Head". Other agents within a holon are called "body-agents". Head can be elected by holon's members, or it can be a predefined agent. The main idea of HOMAS is to assign a task which a single agent cannot accomplish to a holon. The holon then decides how to perform

Fig. 1 Holonic multi-agent
systems structure, courtesy
of Botti and Giret [5]

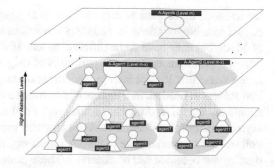

this task. In this way a big/super goal divides to several sub-goals, and each holon is responsible for one or more sub-goals. In the same manner, these goals can be divided into several sub-goals which a holon assigns to its sub-holons. This makes HOMAS design more flexible and simple. The capability of the resulting MAS is an emergent functionality that may surpass the capabilities of each individual agent [16]. Often, problems are neither completely decomposable nor completely non-decomposable. In many hybrid cases, some aspects of the problem can be decomposed, while others cannot.

Holonic agents are structured hierarchically. They can easily realize actions of different granularity, they are autonomous to a certain degree, and they are pro-active; hence holonic agents systems can naturally deal with problems of that type [11]. Agents acting in such structures can encapsulate the complexity of subsystems (simplifying representation and design) and modularize its functionality (providing the basis for rapid development and incremental deployment) [16]. Especially, for extra large systems which contain many agents -for example, a simulator of a city's pedestrians or cars- HOMAS efficiency becomes more visible. Holonic dynamism makes these systems able to have better performance in dynamic and complex environments.

The key property of holonic agents is **bounded autonomy**. Agents in HOMAS are rational and self-interested and decide what to do based on their own preferences like general agents in MAS. But, from the time an agent joins a holon, it cannot do whatever it wants. It should obey holon's commitments after this. This does not mean that every action which the agent makes is determined by the holon's Head like fully cooperative MAS. An agent usually can decide when to join a holon, or leave the holon [4].

3   Negotiations in Holonic Multi-agent Systems

In HOMAS, agents need to coordinate or reach agreement during their activities like other MAS. Generally, there are several ways to reach an agreement in MAS. One of the most common ways is negotiation. In a negotiation process, two or more parties try to reach agreement as soon as possible. They usually want to maximize their payoff too. Negotiations in HOMAS have common properties and characteristics of MAS, but they have some differences too. If the overall problem is decomposed into sub-problems that are not partitions of original one, but there is some overlap in the sense that logical interdependencies occur, communication among the problem solvers is needed. Sub-agents of a holon are communicative and hence, holonic agents are useful in domains of this type. Furthermore, a domain often induces an asymmetric communication behavior between problem solvers in the sense that each unit does not communicate to all other units equally often, i.e., patterns in the communication behavior can be observed. These patterns indicate possible structures for holonic agents: Holons provide facilities for efficient intra-holonic communication, supporting higher frequent communication inside the holon than among different holons (inter-holonic) [11].

The connections (among the agents) in HOMAS can be within a holon or between the holons. Head is responsible for the connections between the holons. In this way, three types of negotiation can be considered in HOMAS. These three types are shown in Fig. 2. The first is among two or more Heads. The second is between a holon's Head and its body-agents, and the last is among the body-agents. The first and third types of negotiation are like general negotiations in MAS. Common mechanisms and settings of negotiations can be used in these negotiations, too. But, the second type is the type which we called *holonic negotiations*. There are a lot of circumstances that a holon's Head decides to reach agreement on something with other holon's agents. For example, in task assigning, Head can negotiate with the agents about what task each agent prefers to accomplish. The main difference of this type of negotiation with other type of negotiations is the possibility of overrule in the negotiation. In general negotiations a self-interested negotiator agent continues the negotiation while it's confident about gaining payoff, and it can also leave the negotiation process whenever it wants. In HOMAS, an agent wants to maximize its utility, but it also knows that it is possible for Head to overrule its decision. Head's overruling means that the agent is forced to do something not based on its preferences, instead because of its commitments to the holon. In this manner, all of negotiation's configurations are affected by these characteristics. Head should decide when to terminate the negotiation and other agents should always consider that if they do not compromise enough, it is possible that they do not gain any utility.

In real negotiations, the main source that agents can obtain useful information about properties of other negotiation parties is the negotiation history [9, 15]. Other assumption about the agents -like knowing preferences or willingness to cooperate

Fig. 2 Three different negotiation types in HOMAS

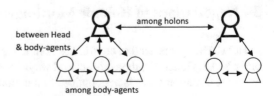

of each other- may be in conflict with the fact that each agent tries not to reveal its privately owned information. So, if an agent wants to learn to improve its negotiation result, it should use the negotiation history. The learning method which is used in holonic environments must be simple and fast. This issue becomes more important when we consider the environments where HOMAS are usually used in. HOMAS usually are used in complex and very large systems which have too many agents. In these systems, holonic structure helps to design the system in more simple and efficient way. In these cases, the system's goal is divided into smaller goals, and each goal is assigned to a holon. Several agents join together and make a holon, and Head decides what each agent should do. In this manner a holonic negotiation scheme should be simple and efficient. It should be fully operable in real time usages.

Other specifications, which make Holonic negotiations distinct, are:

- In HOMAS, an agent can be body-agent of a holon and Head of another holon in the same time. According to this, the agent should always consider its role in the negotiation process in order to choose proper strategy or utility function. As a Head, an agent has totally different responsibilities from when it negotiates as a body-agent.
- When Head negotiates with the holon's body-agents, it may encounter similar agents which have similar properties and negotiation style or they may be dissimilar agents. In HOAMS terms, Head may be in a homogeneous holon or in a heterogeneous holon. In the case where the agents have distinct properties, Head should learn different negotiation styles/strategies.
- Another issue about Head is that Head has such a utility function which has direct relation to the holon's utility. In other words, Head's utility increases when the utility of holon increases. Head tries to increase holon's payoff as a whole. According to this introduction we introduce our method for negotiations in HOMAS.

Among different negotiation protocols available within the multi-agent systems community, some of them could be considered similar to the characteristics of HOMAS negotiations. Specially, negotiation methods that are designed for hierarchical domains are very similar to the idea of holonic negotiations [6, 14].

4 A Specific Mechanism for Negotiations in HOMAS

Here, a special negotiation framework for HOMAS is proposed. In this framework, whenever a negotiation process starts, body-agents propose their offers in every round and Head checks the proposed offers and if the agreement criterion was met, it will inform others about the agreement.

Head and body-agents both try to learn. Linear regression is the learning method which is used in our method. Linear regression is a simple and powerful method which can efficiently be used in real time and dynamic usages. Determining proper independent variables (terms whose values are known) which affects dependent variables (terms which should be predicted) is too important in the regression learning ability. Head wants to know how many rounds the negotiation will last, or when the negotiation will reach to agreement. This is because Head cannot allow body-agents to negotiate for unlimited period. In the other hand, a body-agent is willing to know what other agents propose in each round. It can use this information in order to decide what offer to propose.

A common point among negotiations in humans or artificial agents is that in order to choose the most appropriate offer to propose, they usually try to predict others' offers. Here, for prediction of other agents' next offers, an agent uses linear regression. The independent variables of regression algorithm are previous offers which other agents proposed in the previous round plus the current round number. Holonic agents have special behavior which makes them simultaneously self-interested and cooperative. This behavior is somehow like the semi-cooperative behavior which was described earlier. Semi-cooperative behavior is implemented in different ways. One of these implementations assumes that the agent tries to maximize its utility until some round, and after that point the agent tries to cooperate. In our method, we assumed that this point is a round called *warning round'*. Warning round is the round which Head decides to terminate the negotiation in several rounds later. In this round, Head tells other agents how many rounds they have before negotiation overrule.

In every round, each body-agent runs a thread of regression algorithm in order to predict other agents' next offer. Then, the agent uses this information to decide what offer to propose in the next round. The details of this decision making will be illustrated later. Head uses the same learning approach to predict when the current negotiation will reach to an agreement. Head records the offers that each agent proposes in every round, and like other body-agents it predicts body-agents' next offer using regression. When the negotiation starts, using this data, it predicts body-agent's offers in two round later and then three and so on. Also, based on this information it predicts the agreement round. After this, Head decides when to inform body-agents about the number of remained rounds which they can reach agreement (announcing warning round). Head makes this decision based on the problem configuration. It mainly depends on the time pressure of the holon's domain. As the time pressure increases, Head decreases the number of remained rounds. Head uses warning round as a tool to force body-agents to compromise more.

During the first runs of negotiation, an agent mainly tries to gather useful information which helps it in learning phase. In order to implement this behavior an exploration probability', P_a, is assigned to each agent. An agent explores the environment with probability of P_a and during this period, it uses a simple greedy approach that only selects the option with maximum utility. As the negotiations proceed, this probability decreases. In addition, every agent has a discount ratio, "Ω". This parameter demonstrates the utility of an offer in the next rounds. Ω is like the Ω parameter in bargaining domains. This parameter has the same effect as *time* in other similar negotiation mechanisms. The value of Ω is between 0 and 1. A greater value of Ω means less importance of time.

The process which a body-agent selects what offer to propose in the next round is different before and after the warning round. Before the warning round, a body-agent firstly sorts all of the offers which it can propose based on its utility function. The offer which maximizes its utility is called as *maximal option* or o_m. The agent will propose this option, if the option guarantees agreement in this round. Otherwise, it checks utility of options which guarantee agreement. The option with maximal utility among these options is called $o_{m'}$. If the utility value of o_m in next round was less than $o_{m'}$ utility in current round, the agent will propose $o_{m'}$. In other words, the agent proposes the option which certainly maximizes its utility.

After the warning round, a body-agent knows that if the agreement is not met within the remained rounds, Head may overrule its decision to all of negotiation's participants. At that time the agent must do something that might have no utility for it. So, to avoid this, body-agents should compromise more. It is logical for this compromise to be proportional to the number of remained rounds, or n. In the warning round, Head tells body-agents how many rounds they still have to reach agreement. The general idea is that the agent firstly selects the maximal option, then the agent checks within the options which guarantees agreement. If there was an option with the utility of equal or greater than $(n'/n * maximal\ option's\ utility + option's\ utility)$ the agent will propose that option (here, n' is number of rounds which are passed after the warning round). In other words, the agent compromise $1/(number\ of\ passed\ rounds)$ of maximal option in each round. Ω has the same effect as previous. Figures 3 and 4 show two examples of the negotiation process for a body-agent before and after warning-round.

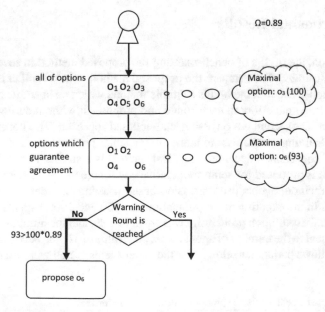

Fig. 3 Warning round is not reached

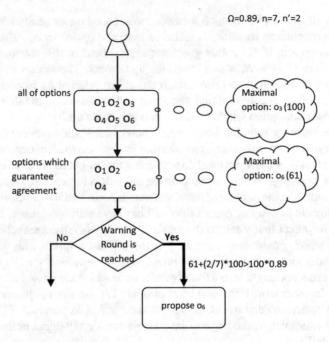

Fig. 4 Two rounds is passed after the warning round, and five rounds is remained until final round

5 Experimental Results

In this section, the results of bench-marking the proposed method in several experiments are illustrated. We compare the proposed method with several existing negotiation methods. The first method is a simple method which an agent just selects the option with maximal utility in each round and proposes it without caring about any other criteria. The second is a Bayesian learner based approach [7]. This method uses a Bayesian learning mechanism to learn other agents' preferences, and use them to make better coordinated decisions. The last method is a similar negotiation mechanism which is proposed for semi-cooperative environments (SC-Ordered-Learner) [8]. In this method, agents use regression based learning in order to learn others' preferences. In the experiments, a population of 1000 and 10000 agents were studied. The initial exploration probability was equal to 0.89, and the number of available tasks was equal to the number of agents. Also, the value of Ω was between 0.7 and 1.

Table 1 shows parameter settings for the experiments which were used.

Table 1 Parameter settings for the proposed method's experiments

Holon's agents	Initial expl. prob.	Num. of tasks	Dis. rate (Ω)
1000–100000	0.89	Equal to agents	Random btw. 0.7 and 1

The designed scenario for the experiments is a task assigning problem. There are a number of tasks which each body-agent should sponsor exactly once. In each round, body-agents propose their offers and Head checks all of these offers. Based on the previous proposed offers, if a permutation was found that all of the tasks could be assigned to the agents, Head announces the agreement. The utility value that agents receive after negotiation and negotiation time are two main factors that agents try to optimize in the negotiation. These two measures have been used in other similar studies in order to show the performance of negotiation mechanisms [2, 3]. Accordingly, in order to compare a negotiation mechanism with other mechanisms, utility and time factors are the basic measures which should be considered.

Figures 5 and 6 show the comparison between negotiation time using the proposed approach and other described methods. Since the negotiations are happening within a consecutive set of rounds, time of negotiation relates to the number of rounds that

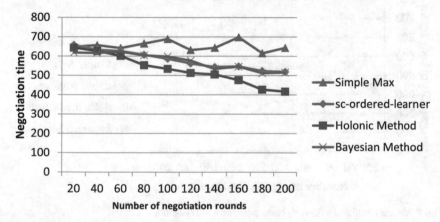

Fig. 5 Average negotiation time for 10000 agents

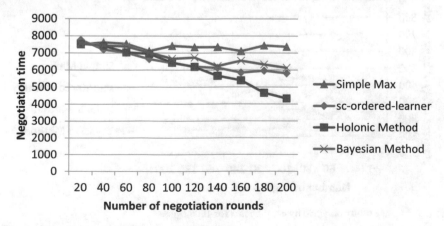

Fig. 6 Average negotiation time for 100000 agents

the negotiation lasts. This is the average number of negotiation rounds. As the figures show, the simple max approach has approximately fixed results, it does not use any learning method and consequently, its performance does not change. The results for other two methods which employ learning algorithms, improve as the number of negotiations increases. The sc-ordered-learner method which seems a better fit to the holonic environments than Bayesian method, has better results than Bayesian method. Another point regarding to this set of results is that as the number of agents increases (from 1000 to 10000) the performance of the proposed holonic method in comparison with other methods increases. We mentioned earlier that HOMAS usually are used in systems which include a very large number of agents, so these results show that the proposed approach can work better in usual Holonic systems environments.

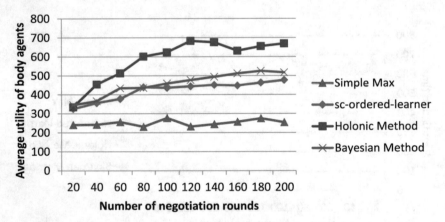

Fig. 7 Average utility obtained by body-agents for 1000 agents

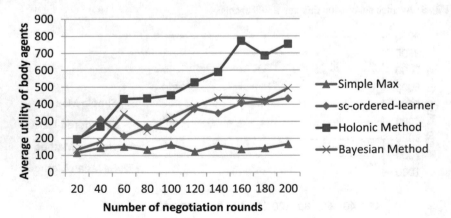

Fig. 8 Average utility obtained by body-agents for 1000 agents

In Figs. 7 and 8, the average utility of all agents involved in the negotiation is shown. This value refers to the average utility value of all of agents after they finished a negotiation with another agent. Like previous results, simple max method did not obtain more than some almost fixed results. Bayesian and sc-ordered-learner methods performance improve when negotiation rounds pass. They approximately have competitive results. Holonic method obtained better results and its results improve as the number of agents increases. Once again, holonic method's performance gets better when the environment becomes larger.

6 Conclusions

In this paper, negotiations in holonic multi-agent systems are studied. The main differences which make this kind of negotiation distinct from general negotiations are illustrated. Most of these differences result from the holonic agents' properties, and others are outcome of the holonic structure. Based on these differences, a specific negotiation mechanism for these domains is proposed. Initial results show that this method can work well in holonic multi-agent systems.

References

1. Arabani, A.B., Ramtin, F., Rafienejad, S.N.: Applying simulated annealing algorithm for cross-docking scheduling. In: Proceedings of the World Congress on Engineering and Computer Science, vol. 2 (2009)
2. Beheshti, R., Mozayani, N.: Predicting opponents offers in multi-agent negotiations using artmap neural network. In: Second International Conference on Future Information Technology and Management Engineering. FITME'09, pp. 600–603. IEEE (2009)
3. Beheshti, R., Mozayani, N.: A new mechanism for negotiations in multi-agent systems based on artmap artificial neural network. In: Agent and Multi-agent Systems: Technologies and Applications, pp. 311–320. Springer (2011)
4. Beheshti, R., Mozayani, N.: HOMAN, a learning based negotiation method for holonic multi-agent systems. J. Intell. Fuzzy Syst. 26(2), 655–666 (2014)
5. Botti, V., Giret, A.: ANEMONA: A Multi-agent Methodology for Holonic Manufacturing Systems. Springer (2008)
6. Bruns, G., Cortes, M.: A hierarchical approach to service negotiation. In: 2011 IEEE International Conference on Web Services (ICWS), pp. 460–467. IEEE (2011)
7. Bui, H., Kieronska, D., Venkatesh, S.: Learning other agents' preferences in multiagent negotiation. In: Proceedings of the National Conference on Artificial Intelligence, pp. 114–119 (1996)
8. Crawford, E., Veloso, M.: Negotiation in semi-cooperative agreement problems. In: IEEE/WIC/ACM Conference on Web Intelligence and Intelligent Agent Technology, vol. 2 (2008)
9. Davami, E., Sukthankar, G.: Online learning of user-specific destination prediction models. In: 2012 International Conference on Social Informatics (SocialInformatics), pp. 40–43. IEEE (2012)
10. Fleyeh, H., Davami, E.: Multiclass adaboost based on an ensemble of binary adaboosts. Am. J. Intell. Syst. 3(2), 57–70 (2013)

11. Gerber, C., Siekmann, J., Vierke, G.: Holonic Multi-agent Systems (1999)
12. Koestler, A.: The ghost in the machine. Psychiatr. Commun. **10**(2), 45 (1968)
13. Lin, F., Lin, Y.: Integrating multi-agent negotiation to resolve constraints in fulfilling supply chain orders. Electron. Commer. Res. Appl. **5**(4), 313–322 (2007)
14. Lin, Y.I., Tien, K.W., Chu, C.H.: Multi-agent hierarchical negotiation based on augmented price schedules decomposition for distributed design. Comput. Ind. **63**(6), 597–609 (2012)
15. Rafinejad, S.N., Ramtin, F., Arabani, A.B.: A new approach to generate rules in genetic algorithm solution to a job shop schedule by fuzzy clustering. In: Proceedings of the World Congress on Engineering and Computer Science, USA (2009)
16. Schillo, M., Fischer, K.: Holonic multiagent systems. KI **17**(4), 54 (2003)
17. Su, Y., Huang, J.: Cooperative output regulation of linear multi-agent systems. IEEE Trans. Autom. Control **57**(4), 1062–1066 (2012)
18. Warnecke, H., Hueser, M., Claypole, M.: The Fractal Company: A Revolution in Corporate Culture. Springer (1993)
19. Wooldridge, M.: An Introduction to Multiagent Systems. Wiley (2009)

Part II
Applications of Automated Negotiations

A Group Task Allocation Strategy in Open and Dynamic Grid Environments

Yan Kong, Minjie Zhang and Dayong Ye

Abstract Against the problem of group task allocation with time constraints in open and dynamic grid environments, this paper proposes a decentralised indicator-based combinatorial auction strategy for group task allocation. In the proposed strategy, both resource providers and consumers are modeled as intelligent agents. All the agents are limited to communicating with their neighbour agents, therefore, the proposed strategy is decentralised. In addition, the proposed strategy allow agents to enter and leave the grid environments freely, and is robust to the dynamism and openness of the grid environments. Tasks in the proposed strategy have deadlines and might need the collaboration of a group of self-interested providers to be executed. The experimental results demonstrate that the proposed strategy outperforms a well-known decentralised task allocation strategy in terms of success rate, individual utility of the involved agents and the speed of task allocation.

Keywords Task allocation · Combinatorial auction · Grid environment

1 Introduction

Task allocation is an important issue to be solved in many domains, such as disaster rescue [6, 12], the RoboCup [13], radar predictions of weather situations [4], and e-trading.[1] In recent years, with the development of the grid environment which is a resource sharing and problem solving environment in virtual organisations [8], much

---

[1]www.ebay.com.

Y. Kong (✉) · M. Zhang · D. Ye
School of Computer Science and Software Engineering,
University of Wollongong, Wollongong 2522, Australia
e-mail: yk573@uowmail.edu.au

M. Zhang
e-mail: minjie@uow.edu.au

D. Ye
e-mail: dayong@uow.edu.au

© Springer International Publishing Switzerland 2016
N. Fukuta et al. (eds.), *Recent Advances in Agent-based Complex
Automated Negotiation*, Studies in Computational Intelligence 638,
DOI 10.1007/978-3-319-30307-9_8

121

attention has been paid to the task allocation in grid environments, in both research [10, 19] and applications [5, 14, 17]. This paper addresses the task allocation in market-based grid environments. Task allocation in market-based grid environments is an important issue because it regulates the resource management and utilisation in the environments.

In the market-based grid environments, resource consumers need the resources of providers to accomplish certain tasks. A task might need the resources provided by more than one provider, and this means that the task needs the collaboration of a group of providers. All the providers and consumers are allowed to freely enter and leave the grid environments, and this represents that the grid environments are dynamic and open. Therefore, the group task allocation strategy should be robust to the dynamism and openness of the grid environments. In addition, when there are more than one candidate provider group that can finish the task, which group should the consumer choose is a challenging issue for the group task allocation. Moreover, task allocation strategies need to be decentralised to guarantee the scalability of grid environment.

A number of group task allocation strategies have been proposed to assemble the multiple resources required by a task, such as auction-based strategies [7, 9], independent negotiations-based strategies [2, 3], greedy algorithm-based strategies [16]. Both of the auction-based strategies in [7, 9], however, need a public auctioneer to run the combinatorial auction for all the consumers. This represents that the strategies in [7, 9] are centralised. In addition, it is difficult to look for an auctioneer which can be trusted by the self-interested bidders in the self-interested grid environments.

The multi-resource negotiation-based strategies proposed in [2, 3] are decentralised. A consumer in their strategies has to negotiate for each required resource type separately. This can cause the heavy communication overhead for the consumer. In addition, the separate negotiations take long time to allocate a task, and this is a disadvantage to a task allocation strategy with time constraints.

The greedy algorithm-based strategy proposed in [16] addresses the problem of group formation for task allocation. In their strategy, the tasks and agents are fixed, not dynamic. Consequently, their strategy is not robust to the dynamism and openness of the grid environments.

In order to meet the challenges and overcome the disadvantages of the current strategies of group task allocation in market-based grid environments, we propose a decentralised Indicator-based Combinatorial Auction Strategy (IAS) for group task allocation in market-based grid environments. In IAS, we introduce an indicator which takes the dynamism and openness of grid environments into account. In addition, there is no central controller in IAS, and an agent is limited to communicating with its neighbours.

The remainder of this paper is organised as follows. The problem overview and description are introduced in Sect. 2. Section 3 introduces the task allocation process and the indicator design, and Sect. 4 experimentally evaluates the proposed strategy. Related work is given in Sect. 5, and we conclude in Sect. 6.

2 Problem Overview and Description

Because each provider might bid for a subset of the resources required by a task, the consumer adopts the combinatorial auction to form candidate groups, from which the consumer picks out the winner one. However, the conventional combinatorial auction-based strategies cannot meet the new challenges of the market-based grid environments any more, and there are two main reasons for this. First, due to the dynamism and openness of the market-based grid environment, there might be frequent contract decommitments from providers, which cannot be handled well by traditional combinatorial auction-based strategies. Thus, we need to adjust the combinatorial auction to make it robust to the openness and dynamism of grid environments. We do this by introducing an indicator for each candidate group to help the consumer pick out the winner one. Second, in the conventional combinatorial auction-based strategy, there is always a public auctioneer which runs the auctions for all the consumers involved in task allocation. However, in the decentralised market-based grid environments, it is hard to find such an auctioneer that is trusted by all the self-interested agents. Therefore, we let each consumer be its own auctioneer in the proposed strategy.

In the sealed first-price combinatorial auction, all bidders submit their sealed bids to the auctioneer, so no bidder knows the bids of other participants. The bidder who provides the highest price is the winner and pays the price it submitted [11]. In the proposed strategy, we adopt the reverse sealed first-price combinatorial auction. It is because in the proposed strategy, the bidder is the provider, and the auctioneer is the consumer. The providers bid to rent their resources to the consumer, but not to buy resources from the consumer, therefore, the winner bidder should be the provider who asks the lowest price for the same resources amongst all the providers. There are mainly two reasons to choose the sealed first-price auction. First, each bidder in the sealed first-price auction has only one chance to submit its bid, and this not only saves time for the whole auction process, but also encourages providers to ask the most reasonable prices in the unique bid. Second, the sealed bid is suitable to the market-based grid environments, because the self-interested agents do not want their private information to be known by other agents.

2.1 Problem Definition

A task in this paper has the latest start time and the deadline, before which the task has to be started and finished respectively. A task might need more than one type of resources that are provided by different providers. Therefore, the providers have to collaboratively accomplish the task and the failure of any provider may result in the failure of the task. A task can be defined as follows:

Definition 1 (Task) A *task*, denoted as τ_k, is defined by a 5-tuple $(R_k, t_{kgen}, d_k, t_{kl}, \bar{r}_k)$, where R_k is the resource set required by τ_k, t_{kgen} is the generation time of τ_k, d_k is the deadline of τ_k, t_{kl} is the latest start execution time of τ_k, and $\bar{r}_k$ represents the maximal reward that τ_k's owner gains if τ_k is finished successfully.

In order to make the resource simple to be described, we assume that a task requires and a provider possesses only one unit of any type of resource at most. Consequently, we will not consider the quantity of resources. Rather, we only consider the number of resource types.

In this paper, each node in the grid environment is modeled as an agent which can make decisions autonomously. When an agent needs other agents' resources to execute its tasks, it is a consumer, and when an agent provides its resources to other agents, it is called a provider. An agent can be a provider or a consumer, or both. Formally, we have:

Definition 2 (*Agent*) An agent, denoted as a_i, is defined by a 3-tuple $(ID_i, R_i, Neigh_i)$, where ID_i (where $ID_i \in \mathbb{N}$) is the unique identifier of a_i, R_i is the resource set that a_i owns, and $Neigh_i = \{a_{i1}, \ldots, a_{ik}\}$ is a_i's neighbour set, where k (where $k \in \mathbb{N}$) is the number of a_i's neighbours.

In order to decentralise the task allocation strategy, each agent has a neighbourhood and is limited to communicating with its neighbours. An agent judges whether its neighbours are still active in the environment through heartbeat messages which can be defined by:

Definition 3 (*Heartbeat Message*) A heartbeat message sent from a_i to a_j, denoted as $HeartBeat_{ij}$, is defined by a 2-tuple (ID_i, ID_j), where ID_i and ID_j are the ID numbers of a_i and a_j, respectively.

In particular, an agent keeps sending a heartbeat message to its neighbours once in each time unit to inform them that it is still active in the environment. If agent a_i did not receive any heartbeat message from its neighbour, say a_j, in the previous time unit, it assumes that a_j has left the environment.

Besides the heartbeat message, there are two other types of messages, i.e., the request message and the reply message for executing tasks, which are defined by Definitions 4 and 5, respectively.

Definition 4 (*Request Message*) A request message sent from a_i to a_j, denoted as $ReqExecute_{ij}$, to request a_j to execute a_i's task is defined by a 4-tuple (ID_i, ID_j, τ_k, HL), where ID_i and ID_j represent the message sender's ID number and receiver's ID number, respectively, τ_k is the task to be executed and HL is a hop limitation (where $HL \in \mathbb{N}$ and $HL \geq 1$), which prevents the request message from being transmitted endlessly.

Definition 5 (*Reply Message*) A reply message sent from a_j to a_i, denoted as $RepExecute_{ji}$, to reply the message $ReqExecute_{ij}$, is defined by a 6-tuple $(ID_j, ID_i, t_{fin}, \tau_k, Set_r, pr_j)$, where ID_j and ID_i represent the message sender's ID number and receiver's ID number, respectively, t_{fin} is the finish time of τ_k offered by a_j (where t_{fin} has to meet the condition of $t_{fin} \leq d_k$, where d_k is the deadline of τ_k), Set_r is the resource set that a_j can provide to τ_k, and pr_j is the asked price by a_j for Set_r.

2.2   Task Allocation Process

When a consumer, say a_i, has a task, say τ_k, to be allocated, a_i sends a request message (see Definition 4) to all of its neighbours. The agent which receives the request message checks its resource set to see whether it has at least one type of the resources required by τ_k. If so, it sends a reply message (see Definition 5) to a_i. If not, and meanwhile HL in the request message is not equals to 0, a_j transmits the request message to its own neighbours. Otherwise, a_j ignores the received message. In order to increase a_j's chances to be selected by consumers, we stipulate that a_j can send reply messages to more than one consumer simultaneously. If a more attractive transaction to a_j appears after a_j already signed a contract with another consumer, a_j is allowed to decommit from the consumer, to whom a_j has to pay a penalty.

If a request message is transmitted, it is re-assembled by the intermediate agents, which means that the intermediate agents will change some information in the request message. The information that is changed by the intermediate agents is the ID of the new receiver of the message. In addition, HL is reduced by 1 every time the request message is transmitted, and the transmitting process will not end until HL becomes 0. If a provider which has at least one type of the required resources of τ_k is not the direct neighbour of a_i, it has to construct neighbourhood relationship with a_i first, and then can straight send a reply message to a_i.

It is possible that a consumer does not receive any reply message in a short time after its request message for executing tasks was sent out, due to the transmission of the request message. The consumer will not start to form the candidate groups until the latest start time of its task arrives. If the consumer has not received any reply message when the latest start time of its task arrives, it will give up the task. Once the consumer starts the group formation, it will ignore any future arriving reply messages. A candidate group is a group of providers which can collaboratively meet the resource requirement of the consumer's task. A candidate group is irreducible, which means that a candidate group cannot finish the task without any provider in this group.

When a consumer finishes the group formation, it starts to do the winner determination. Both the group formation and winner determination will be described in detail in Sect. 3. Each of the providers in the winner group signs a contract with the consumer, and the contract is defined as:

Definition 6 (*Contract*) A contract between consumer a_i and provider group gro (where gro is a set that contains n providers, i.e., $gro = \{a_1, a_2, \ldots, a_n\}$), is a $(5+n)$-tuple $(a_i, gro, \tau_k, rew, t_{fin}, pri_1, pri_2, \ldots, pri_n)$, where rew is the reward that a_i will gain if τ_k is finished before time t_{fin}, t_{fin} is the finish time of τ_k offered by gro (t_{fin} will be particularly formulated in Sect. 3.2.1), and $pri_1, pri_2, \ldots, pri_n$ are the prices that a_i will pay to $a_1, a_2, \ldots, a_n$, respectively, if τ_k is finished no later than t_{fin}.

3 The Group Formation and Winner Determination

In this section, we will in particular describe the group formation process in Sect. 3.1, and introduce the design of the indicator in Sect. 3.2.

3.1 The Group Formation

Suppose that agent a_i sends a request message (ID_i, ID, τ_k, HL) $(\tau_k = (R_k, t_{kgen}, d_k, t_{kl}, \bar{r}_k))$ to its neighbours a_j. We assume that resource $r \in R_k$, and the number of bidders which can provide r is denoted as n_p^r. We use P to denote the bidder set of the bidders which bid for at least one of the required resources of τ_k. There are four steps to form all the candidate groups (the situation under which a consumer starts the group formation has been analysed in Sect. 2.2).

- Step 1: Pick out resource r, where $r = \mathbf{arg}\, min_{r \in R_k}(n_p^r)$.
- Step 2: Pick out the bidders set P_r in which every bidder bids for r, that is, if the provider in P_r at index k is denoted by $P_r[k]$, then $\forall k \in (1, \ldots, |P_r|), r \in R_k$.
- Step 3: For each $P_r[k] \in P_r$, pick out the group $P_{-r}[k]$ from $P \setminus P_r[k]$. If the element of $P_{-r}[k]$ at index j is denoted by $P_{-r}[k][j]$, then $R_k \subseteq (R_{P_r[k]} \cup R_{P_{-r}[k]})$, where $R_{P_{-r}[k]} = \cup_{j=1}^{|P_{-r}[k]|} R_{P_{-r}[k][j]}$ and $\forall h \in (1, \ldots, |P_{-r}[k]|)$, $R_{P_r[k]} \cup (R_{P_{-r}[k]} \setminus R_{P_{-r}[k][h]}) \subset R_k$. a_i obtains $P_{-r}[k]$ from $P_r \setminus P_r[k]$ through recursively calling steps 1, 2 and 3. We do the same to each $P_r[i] \in P_r$ $(i \in 1, \ldots, |P_r|)$ as what we do to $P_r[k]$.
- Step 4: Suppose that from steps 1, 2 and 3, a_i can obtain n candidate groups for τ_k in total, and we use C_k to denote the set of the candidate groups. We denote the element of C_k at index i $(1 \le i \le |C_k|)$ by $C_k[i]$. What we do in this step is to delete group $C_k[i]$, if $\exists j \in (1, \ldots, |C_k|)$ and $j \ne i$, $C_k[j] \subseteq C_k[i]$.

Now we assume that the consumer obtained k candidate groups, gro_1, gro_2, ..., gro_k. The indicators of these groups are $indic_1, indic_2, \ldots, indic_k$ (the indicator will be introduced in detail in the next section), respectively, and the prices asked by these groups are $pri_1, pri_2, \ldots, pri_k$, respectively. We stipulate that the winner group is the one whose ratio value between the price asked by the group and the indicator of the group is the smallest. From the above analysis, the winner group is gro_j, where

$$j = \mathbf{arg}min_i f(i), \tag{1}$$

where $1 \le i \le k, 1 \le j \le k$ $(i, j \in \mathbb{N})$, and

$$f(i) = pri_i / indic_i \tag{2}$$

When there are more than one group that can meet Eq. (1), a_i picks out the winner randomly from the groups. The indicator design and the reason to decide the winner group according to Eqs. (1) and (2) will be introduced next.

3.2 The Indicator Design

In this section, we first introduce and formulate the factors that should be taken into account by the indicator in order to make the task allocation strategy robust to the dynamism and openness of the grid environment, and then we accordingly formulate the indicator.

3.2.1 Consideration of Factors for the Indicator Design

In addition to the price asked by a candidate group, there are four other factors taken into account when a consumer picks out the winner group in the proposed strategy. The four factors are (i) the probability of no decommitment in a group, (ii) the number of overlapping resources in a group, (iii) the reputation of a group, and (iv) the reward a consumer can gain from a group.

(i) *the probability of no decommitment.*

The likelihood of external offers and opportunities may provide incentives to providers to deviate from commitments [15]. In order to make the proposed task allocation strategy suitable to the dynamism and openness of the environment, all of the providers are allowed to decommit from contracts. Because each group is irreducible, the decommitment of even one provider may result in the failure of the task unless another provider joins the group to replace the decommited one in time. However, even though the replacement is possible, there is no guarantee that the fill-in will be found in time before the latest start time of the task. Therefore, the lower the decommitment probability, the less chance of the failure of a task. Consequently, the decommitment probability of a candidate group is one of the key factors to be considered by a consumer when the consumer determines the winner group.

It is reasonable to assume that different providers have different probabilities to decommit from a contract, and the decommitments of providers are independent, that is, a provider's decommitment is an independent decision making, not affected by any other providers in the same group. Many reasons can induce a provider to decommit from a contract, such as a more attractive task available, an accident.

In the proposed strategy, a consumer predicts the probability that a provider will not decommit with it according to the trade history between itself and the provider. Suppose that consumer a_i and provider a_j already trade with each other for k_1 ($k_1 \in \mathbb{N}$) times before, and a_j decommitted with a_i for k_2 ($k_2 \in \mathbb{N}$)

times in total. The probability predicted by a_i that a_j will not decommit with a_i is denoted by $p_{nodec}(j)$, where

$$p_{nodec}(j) = e^{-k_2/k_1}\ (k_1 \geq 1) \tag{3}$$

From Eq. (3), we have $1/e \leq p_{nodec}(j) \leq 1$.

There are two main reasons for us to formulate $p_{nodec}(j)$ in Eq. (3). The first reason is that the larger k_2/k_1 is, the smaller $p_{nodec}(j)$ should be. The other reason is that we should avoid that $p_{nodec}(j)$ is equal to 0, because being equal to 0 of any $p_{nodec}(j)$ in a provider group will be dominating (the reason for this will be seen later in Eq. (5)). Consequently, we employ the exponential function of 'e' to meet the two conditions.

If $k_1 = 0$, a_i assigns $(1 - 1/e)/2$ to p_{nodec}. It is because $(1 - 1/e)/2$ is the median between $1/e$ and 1, and this is reasonable because a_i and a_j have never trade with each other before. In summary, we have:

$$p_{nodec}(j) = \begin{cases} e^{-k_2/k_1} & (k_1 \geq 1) \\ (1 - 1/e)/2 & (k_1 = 0) \end{cases} \tag{4}$$

Now we assume that consumer a_i already signed a contract provider group gro ($gro = \{a_1, a_2, \ldots, a_n\}$). The probabilities of the n providers not to decommit with a_i are $p_{nodec}(1), p_{nodec}(2), \ldots, p_{nodec}(n)$ ($p_{nodec}(i) \in \mathbb{R}, 1 \leq i \leq n$), respectively. If we use $P_{nodec}(gro)$ to denote the probability that no provider in gro will decommit with a_i, then according to the multiplication principle, we have:

$$P_{nodec}(gro) = \prod_{j=1}^{n} p_{nodec}(j) \tag{5}$$

Apparently, $0 < P_{nodec}(gro) \leq 1$. From Eq. (5), we know that being equal to 0 of any $p_{nodec}(j)$ will result in $P_{nodec}(gro)$ being equal to 0. Therefore, being equal to 0 of any $p_{nodec}(j)$ will put dominating impact on $P_{nodec}(gro)$, and this is the reason for us to formulate $p_{nodec}(j)$ in Eq. (3) when $k_1 \geq 1$ to avoid that $p_{nodec}(j)$ is 0.

(ii) *the number of overlapping resource types in a group*

There might be overlapping resources among the resources provided by all the providers in a group. Suppose that there are only a_i and a_j in provider group gro. The resource set that a_i can provide is denoted by Set_i, and $Set_i = \{1, 2, 3\}$, where '1' represents resource type 1, and so on. The resource set that provided by a_j is Set_j, and $Set_j = \{3, 4, 5\}$. The number of overlapping resource types in gro is 1, and the overlapping one is resource type '3'. Suppose that there are n providers in gro, which are $a_1, a_2, \ldots, a_n$. When we calculate the number of overlapping resource types of gro, we first calculate the number of overlapping resource types amongst a_1 and a_2. Then we treat a_1 and a_2 as one entirety, say, a_{12}. We can use the same way to calculate the number of overlapping resource types amongst a_{12} and a_3, and so on. Finally, the accumulated number of overlapping resource types is the number of overlapping resource types in gro.

Intuitively, the more overlapping resource types in a group, the more waste of the overlapping resources. It is because in our strategy, we assume that a task only requires one unit of any resource type. Therefore, few or even no overlapping resource types in a group is desirable.

(iii) *the reputation of a group*

The QoS (Quality of Service) of providers may be different, so it is important for consumers to take the potential QoS of providers into account when choosing the winner provider group. To address this concern, consumers anticipate the potential QoS of providers according to the providers' past QoS.

Based on the trust model proposed by Yu and Singh [18], we propose a general method for consumers to formulate providers' reputations in terms of QoS. A consumer can evaluate the reputation of a provider according to different criteria, depending on the specific situations. For example, if tasks are urgent, and very time-sensitive, a consumer can evaluate the reputation of a provider based on whether the provider finished tasks before deadlines in the past services.

Assume that consumer a_m wants to predict the reputation of provider a_n. a_m has the latest k $(k \geq 1)$ services from a_n, and the k services are denoted as $S = \{s_1, s_2, \ldots, s_k\}$. We use $r_1, r_2, \ldots, r_k$ to represent the reputation values of $s_1, s_2, \ldots, s_k$ respectively, and $r_i \in (0, 0.1, 0.2, \ldots, 1)$ $(1 \leq i \leq k)$. We define a low threshold ω $(0 < \omega < 1)$ and an upper threshold Ω $(0 < \omega < \Omega < 1)$ for the QoS. $S(T)$ is the support set of trust, and $S(T) = \{s_i | \Omega \leq r_i \leq 1, 1 \leq i \leq k\}$. $S(\neg T)$ is the support set of distrust, and $S(\neg T) = \{s_j | 0 \leq r_j \leq \omega, 1 \leq j \leq k\}$. When $k \geq 1$, we formulate the reputation of a_n which is denoted as r_{a_n} by:

$$r_{a_n} = e^{(|S(T)| - |S(\neg T)|)/k} \tag{6}$$

From Eq. (6), we know that $\frac{1}{e} \leq r_{a_n} \leq e$. The reason to formulate r_{a_n} in Eq. (6) is to guarantee that the higher $|S(T)| - |S(\neg T)|$ is, the higher r_{a_n} is.

When $k = 0$, which means that a_m has never got service from a_n, it is reasonable for a_m to assign $(e - \frac{1}{e})/2$ to r_{a_n}, because $(e - \frac{1}{e})/2$ is the median of r_{a_n}'s value range $[\frac{1}{e}, e]$.

From the above analysis, we can formulate the reputation of a_n by:

$$r_{a_n} = \begin{cases} e^{(|S(T)| - |S(\neg T)|)/k} & (k \geq 1) \\ (e - \frac{1}{e})/2 & (k = 0) \end{cases} \tag{7}$$

Now we assume that the provider group $gro = \{a_1, a_2, \ldots, a_n\}$, and the reputations of the n providers in gro are $r_{a_1}, r_{a_2}, \ldots, r_{a_n}$, respectively. If $r_{min} = min\{r_{a_1}, r_{a_2}, \ldots, r_{a_n}\}$, and $r_{max} = max\{r_{a_1}, r_{a_2}, \ldots, r_{a_n}\}$, then the reputation of gro is formulated as:

$$r_{gro} = \begin{cases} ((\sum_{i=1}^{n} r_{a_i}) - r_{min} - r_{max})/(n-2) & (n \geq 3) \\ (\sum_{i=1}^{n} r_{a_i})/n & (1 \leq n \leq 2) \end{cases} \quad (8)$$

The reason to formulate r_{gro} in Eq. (8) is to avoid the dominating impacts of the maximal and minimal reputations on gro when $n \geq 3$.

(iv) *the reward gained by a consumer*

In the proposed strategy, consumers can gain rewards if and only if their tasks are finished before task deadlines. The earlier a task is finished, the more rewards the consumer can gain. Because a task needs the collaboration of the providers in a group, the task cannot be finished until all of the providers finish their jobs. Consequently, if the finishing times of all the providers in the provider group are $t_{fin1}, t_{fin2}, \ldots, t_{fink}$ (k is the number of providers in the group), then the finishing time of the task denoted by t_{fin} is: $t_{fin} = max\{t_{fin1}, t_{fin2}, \ldots, t_{fink}\}$. We adopt a piecewise function to represent the relationship between the finishing time of task τ_k, say t, and the reward that τ_k's owner can gain when τ_k is finished, say $rew_k(t)$. Without loss of generality, we formulate $rew_k(t)$ by:

$$rew_k(t) = \begin{cases} \overline{r}_k/(t_1 - t_{kgen}) & t_{kgen} < t \leq t_1 \\ \cdots \\ \overline{r}_k/(t_{i+1} - t_{kgen}) & t_i < t \leq t_{(i+1)} \\ \cdots \\ \overline{r}_k/(d_k - t_{kgen}) & t_n < t \leq d_k \\ 0 & d_k < t \end{cases} \quad (9)$$

where $\overline{r}_k$ is the maximal reward that task τ_k's owner can gain when τ_k is finished, t_{kgen} and d_k are the generation time and the deadline of τ_k, respectively (see Definition 1). $t_1, \ldots, t_i, \ldots, t_n$ are different times, and $t_{kgen} < t_1 < \cdots < t_i \cdots < t_n < d_k$.

The reason to formulate $rew_k(t)$ in Eq. (9) is to reflect that the bigger t is, the smaller $rew_k(t)$ is, i.e., the later τ_k is finished, the fewer reward τ_k's owner can gain.

3.2.2 Formulation of the Indicator

In order to enable consumers to evaluate the candidate groups according to the four factors analysed above, we formulate an indicator for each candidate group. The indicator comprehensively takes the four factors into account. Suppose that gro is one of the candidate groups for task τ_k, and a_i is the owner of τ_k. $No.r$ is the accumulated number of overlapping resource types in gro; the probability that no provider in gro will decommit with a_i is P_{gro}; the time when gro can finish τ_k is t_{fin}; $rew(t_{fin})$ is the reward that the consumer can gain if τ_k is finished at time t_{fin}; $\bar{r}$ is the maximal reward that a_i can gain when τ_k is finished, and the reputation of gro is r_{gro}. Then the indicator of gro which is denoted by $indic_{gro}$ is formulated as:

$$indic_{gro} = P_{gro} + 1/log_2(No.r + 2) + rew(t_{fin})/\bar{r} + r_{gro} \qquad (10)$$

Because $0 \leq rew(t_{fin})/\bar{r} \leq 1$, $1/e \leq r_{gro} \leq e$, $0 \leq P_{gro} \leq 1$, and $0 \leq No.r \leq \infty$, it is apparent that even the small fluctuation of $No.r$ will be dominating to the value of $indic_{gro}$. In order to avoid the dominating impact of $No.r$ on $indic_{gro}$, we restrain the impact of $No.r$ by introducing "log_2" into Eq. (10). Additionally, in order to avoid the situations of "$1/log_2 0$" and "$1/log_2 1$" when $No.r$ is equal to 0 and 1, respectively, we introduce "$1/log_2(No.r + 2)$" instead of "$1/log_2(No.r)$" into Eq. (10).

4 Experiment

In order to experimentally evaluate the performance of the proposed strategy in this paper, we compare it with the task allocation strategy proposed in [3], which is denoted as MRN (Multi-Resource Negotiation).

In MRN, a consumer allocates its task through negotiating with providers, and negotiates with providers for each of the required resource types separately. In the proposed strategy in this paper which is denoted by IAS (Indicator-based Task Allocation Strategy), the consumer adopts an indicator-based combinatorial auction strategy to allocate tasks. The reason to choose MRN to be the experimental benchmark is that MRN is a classical and well-known task allocation strategy in recent years.

4.1 Experimental Criteria

One of the main purposes of task allocation strategyes in grid environments is to successfully allocate as many tasks as possible [8]. Thus, in the experiment, we study the success rate of task allocation. The utility of the agent involved in task allocation is also important to the performance of a task allocation strategy [2]. Additionally, because the agents in IAS are self-interested, we test the individual utility of the agents involved in task allocation. However, there are too many agents and thus, it is impossible to show the utility of each agent, consequently, we report the utility distributions of all the agents involved in the task allocation. We also study the total time used for the whole task allocation process to test the speed of IAS. It is because when tasks have deadlines, speed is important for a task allocation strategy. In summary, in the experiment, we test three main criteria: the success rate of task allocation, the utility distribution of agents involved in task allocation (the utility of an agent will be in particular described next), and the total time used for the whole task allocation process.

4.1.1 Formulation of Utility

(1) The utility that agent a_i gains from task τ_k is formulated as:

$$uti_{ik} = (rew_{ik} - cost_{ik})/rew_{ik}, \tag{11}$$

where rew_{ik} is the reward that a_i can gain when τ_k is finished, and $cost_{ik}$ is the cost of a_i to finish τ_k. rew_{ik} and $cost_{ik}$ are specified in two cases:

  (i) If a_i is the owner of τ_k, $cost_{ik}$ is the price a_i pays to the provider group that finished τ_k.
 (ii) If a_i is one of the providers in the group which finished τ_k, we have:

$$cost_{ik} = costres_{ik} + pel_{ik}, \tag{12}$$

where $costres_{ik}$ is the cost of a_i's resources to finish τ_k, and pel_{ik} is the penalty that a_i pays to other consumers with which a_i decommitted in order to execute τ_k (the penalty will be introduced in Sect. 4.1.2).

(2) The utility that group gro gains from τ_k is the sum of all the utilities gained from τ_k by all the providers in gro:

$$uti_{grok} = \Sigma_{i=1}^{n} uti_{ik}, \tag{13}$$

where n is the number of providers in gro.

4.1.2 Formulation of Penalty

When a provider decommits from a contract, there is at least one victim agent. A victim agent is an agent that encounters the risk of loss of rewards due to the decommitment by the decommitting agent. For example, when a provider, say a_j, decommits from a contract with consumer a_i, then a_i risks not being able to find another provider to replace a_j before the latest start time of its task. Besides, if there are other providers in the same provider group with a_j, these providers also encounter the risk of gaining no rewards if the task fails. In this situation, a_i and the other providers in the same provider group with a_j are victim agents. The decommitting providers are required to pay penalties to the corresponding victim agents.

According to Definition 6, con $(a_i, gro, \tau_k, rew, t_{fin}, pri_1, pri_2, \ldots, pri_n)$ is a contract between consumer a_i and group gro. Suppose that a_p and a_q are two of the providers in *gro*, if a_p decommits from *con*, both a_i and a_q are victims and thus, a_p has to pay penalty to them.

(1) The penalty that a_p pays to a_i is formulated as:

$$pel_{pi} = \beta \times rew \tag{14}$$

(2) The penalty that a_p pays to a_q is formulated as:

$$pel_{pq} = \beta \times pri_q \tag{15}$$

where $0 < \beta < 1$.

Due to the time-sensitivity of tasks, in the proposed strategy in this paper, β is formulated by:

$$\beta = (t - t_{kgen})/(d_k - t_{kgen}) \tag{16}$$

where t is the time when the decommitment happens, t_{kgen} and d_k are the generation time and deadline of τ_k, respectively.

Intuitively, Eqs. (14), (15) and (16) represent that the later a provider decommits from a contract, the more penalty it should pay to the corresponding victim agents.

4.2 Experimental Settings

We simulate the grid environment by generating 100 agents and 300 tasks before the experiment starts. The 300 tasks are distributed to the 100 agents randomly. An agent is a consumer if at least one task is assigned to it. Each agent has at least one type of resource, thus, every agent can be a provider. We activate 50 of the 100 agents before the experiment starts, and the other 50 agents will be activated at a fixed probability after the experiment starts. After the start of the experiment, in each time unit, each agent in the environment leaves the environment at 0.2 probability, and each agent out of the environment can enter the environment at 0.3 probability. When an agent newly enters into the environment, it has a 0.1 probability to become neighbours with any agent in the environment. In the experiment, we test the performance of IAS in two scenarios.

4.2.1 Scenario 1: Examination of the Impacts of the Required Resource Type Number per Task

In Scenario 1, we test the impacts of the required resource type number per task on the success rate of task allocation, the utility distribution of the agents involved

Table 1 Parameter settings for scenario 1

Variables	Meanings	Values
N_{pro}	The number of provided resource types per provider	[1, 10]
N_t	The number of tasks in the environment	[0, 300]
$flex$	The available time to allocate a task, i.e., the allocation flexibility	(0, 10] (s)

in task allocation, and the total used time for the whole task allocation process. The parameter settings for Scenario 1 are listed in Table 1.

From Table 1, it can be seen that the number of resource types provided by each provider is in-between 1 and 10. The reason to choose the range of [1, 10] is that in real life applications, it is common for each provider to provide from 1 to 10 types of resources. The number of tasks in the grid environment does not remain constant. Rather, it varies over [0, 300].

We use *allocation flexibility* which is denoted as $flex$ to represent the remaining time for a consumer to allocate its task before the task's latest start execution time. If t_{kl} is the latest start execution time of task τ_k, and t is the current time, apparently, the allocation flexibility of τ_k denoted by $flex_k$ is:

$$flex_k = t_{kl} - t \qquad (17)$$

From Table 1, we can see that we set a range of values for each parameter, and a parameter takes different values randomly over the range at different time units during the course of the experiment. The reason for this is to reflect the dynamism and openness of grid environments.

4.2.2 Scenario 2: Examination of the Impacts of Allocation Flexibility

In Scenario 2, we test the impacts of allocation flexibility which is formulated by Eq. (17) on the success rate of task allocation, the utility distribution of the agents involved in task allocation, and the total time used for the whole task allocation process. The parameter settings for Scenario 2 are listed in Table 2.

Similar to the parameter settings for Scenario 1, we set a range for each parameter in Scenario 2 to reflect the dynamism and openness of the grid environment.

Table 2 Parameter settings for scenario 2

Variables	Meanings	Values
N_{req}	The number of required resource types per task	[1, 10]
N_{pro}	The number of provided resource types per provider	[1, 10]
N_t	The number of tasks in the environment	[0, 300]

The number of required resource types per task varies between 1 and 10, the number of provided resource types per provider varies in-between [1, 10], and the number of tasks in the grid environment changes over time during the course of the experiment, moving any value in-between 0 and 300.

4.3 Experimental Results and Analysis

4.3.1 Results and Analysis of Scenario 1

Figure 1a shows the impact of the number of required resource types per task (N_{req}) on the success rate of task allocation. From Fig. 1a, it can be seen that when N_{req} varies from 1 to 10, the success rate of IAS is always higher than that of MRN. The reason is that when the consumers in MRN choose the providers, they do not take the decommitment probabilities of the providers into account. This might cause the high decommitment probability of the chosen providers, which will directly result in high probability of tasks' failures.

Figure 1b illustrates the impact of N_{req} on the total time used by the whole task allocation process. It can be seen from Fig. 1b that the time used by IAS is shorter than that of MRN, when N_{req} varies from 1 to 10. This is because in MRN, a consumer has to negotiate with providers for each required resource separately. Additionally, each negotiation may contains more than one round of bargaining. IAS adopts the

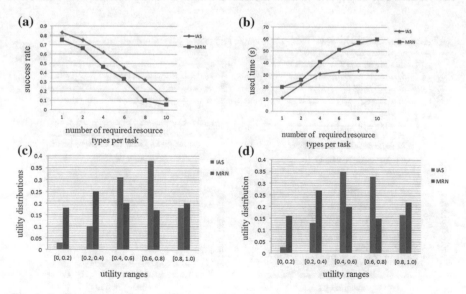

Fig. 1 Performance based on different number of required resource types per task. **c** Average number of required resource of types: 4, **d** Average number of required resource of types: 6

sealed first-price combinatorial auction, which can save time as compared with the
separate multi-resources negotiation in MRN. For this reason, the time used by IAS
is shorter than that by MRN.

Figure 1c, d show the utility distributions when N_{req} is 4 and 6, respectively. From
Fig. 1c, d, we can see that more utilities of agents in IAS fall into the middle ranges
(e.g., [0.4, 0.6) and [0.6, 0.8)) as compared with MRN. This is because in IAS, when
a consumer picks out providers to execute its tasks, in addition to the prices asked by
the providers, the consumer takes decommitment probability into account. However,
the consumer in MRN only considers the prices asked by providers when picking out
providers. In addition, the separate negotiation for each required resource type always
results in there being more providers eventually chosen to execute a task in MRN,
as compared with IAS. This always causes the higher decommitment probability in
MRN than that in IAS. In both IAS and MRN, if a task fails, the utility that the
consumer can gain from the task is equal to 0. Consequently, compared with IAS,
the higher decommitment probability results in more utilities of consumers in MRN
falling into the extreme ranges (e.g., [0, 0.2)).

4.3.2 Results and Analysis of Scenario 2

Figure 2a shows the success rates of task allocation based on different allocation
flexibilities ($flex$). We can see from Fig. 2a that when $flex$ varies in the range of
[0, 10] and the range of [50, 60], the success rate of IAS is always higher than that of
MRN. This can be explained from the viewpoint of task allocation speed. In MRN,

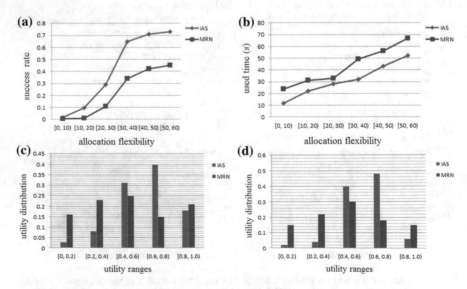

Fig. 2 Performance based on different allocation flexibilities of tasks. **c** Allocation flexibility:
30–40, **b** Allocation flexibility: 40–50

the task allocation process contains the negotiations for each required resource type, and each negotiation may include more than one round of bargaining. Therefore, compared with the sealed first-price combinatorial auction adopted by IAS, MRN needs longer time to allocate a task. When tasks have time constraints, the speed of a task allocation strategy is one of the key factors that affect the success rate of task allocation. For this reason, the success rate of IAS is higher than that of MRN.

Figure 2b presents the times used by the whole task allocation process based on $flex$. As can be seen from Fig. 2b, the time used by IAS is shorter than that of MRN. The reason for this is the faster speed for task allocation in IAS than that in MRN, and this has been analysed in Fig. 2a.

Figure 2c, d show the utility distributions over the range of [0, 1] when $flex$ is chosen from the range of [30, 40) and the range of [40, 50), respectively. From Fig. 2c, d, we can see that more utilities of agents in IAS fall into the middle ranges (e.g., [30, 40), [40, 50)) as compared with MRN. The reason for this is the same as that in Fig. 1c, d. In addition, we can see from Fig. 2c, d that when $flex$ is chosen from the range of [40, 50), more utilities of agents in both MRN and IAS fall into the middle ranges, compared with the situation that $flex$ is chosen from the range of [30, 40). This is because when the allocation flexibility is larger, consumers in both MRN and IAS have longer time to look for providers to replace the de-committed ones. Additionally, the consumers in MRN have longer time to negotiate. For this reason, the probability of the failures of tasks can be decreased. The smaller probability of the failures of tasks results in more utilities falling into the middle ranges.

5 Related Work

In [3], An et al. proposed a negotiation-based strategy which is called multi-resource negotiation to allocate tasks. In their strategy, a consumer negotiates with providers for each of its required resources separately. However, separate negotiation always causes a large number of providers be chosen to finish a task, and too many providers in a group to execute a task may result in a communication overload among these providers. Moreover, separate negotiation for each required resource might make the time needed by successfully allocating a task too long, and this is a disadvantage to urgent tasks.

Kraus and Shehory proposed an auction-based coalition formation strategy for task allocation [9]. When there are more than one coalitions which can finish a task, the auctioneer chooses the winner only according to the bids submitted by these coalitions. Thus, compared with their strategy, the proposed strategy in this paper gives greater consideration to the dynamism and openness of the environment. In addition, in their strategy, the auctioneer is a central controller, while in the proposed strategy in this paper, each consumer is its own auctioneer, and this not only decentralises the task allocation, but also improves the management of self-interested agents' privacy.

A reinforcement learning-based coalition formation strategy was proposed by Abdallah and Lesser [1]. They introduced reinforcement learning into their strategy to help a consumer choose the proper group of providers for a task. In their strategy, once a provider joins a group, it is not allowed to leave to join any other group. If some urgent tasks come into the environments, the already-chosen providers are not allowed to be transferred to these new tasks, and this may result in the failure of urgent tasks. Thus, the strategy in [1] is not suitable to the dynamic and open environments.

K.S. Macarthur et al. proposed a branch-and-bound fast-max-sum, decentralised, dynamic task allocation algorithm [10]. In their algorithm, the grid environment is dynamic, which means that both agents and tasks are dynamic and change over time. However, in their algorithm, each agent tries to maximise the global utility cooperatively, while neglecting its own utility.

6 Conclusion

This paper proposed an indicator-based combinatorial auction strategy for group task allocation in open and dynamic grid environments. The proposed strategy addressed the challenges of decentralisation, the dynamism and openness in the grid environments, by introducing indicators to help consumers choose the winning provider groups. The experimental results illustrate that the proposed strategy outperforms a well-known strategy in term of success rate of task allocation, average utility of involved agent, and the speed of task allocation. In future work, we intend to extend this work to task allocation when a task contains more than one subtasks with multiple constraints, such as preference constraints, resource constraints, etc.

References

1. Abdallah, S., Lesser, V.: Organization-based cooperative coalition formation. In: International Conference on Intelligent Agent Technology, pp. 162–168 (2004)
2. An, B., Lesser, V., Irwin, D., Zink, M.: Automated negotiation with decommitment for dynamic resource allocation in cloud computing. In: Proceedings of AAMAS, pp. 981–988 (2010)
3. An, B., Lesser, V., Sim, K.M.: Strategic agents for multi-resource negotiation. Auton. Agents Multi-Agent Syst. **23**(1), 114–153 (2011)
4. An, B., Lesser, V., Westbrook, D., Zink, M.: Agent-mediated multi-step optimization for resource allocation in distributed sensor networks. In: AAMAS, pp. 609–616 (2011)
5. Buyya, R., Abramson, D., Venugopal, S.: The grid economy. Proc. IEEE **93**(3), 698–714 (2005)
6. Chapman, A., Micillo, R., Kota, R., Jennings, N.: Decentralised dynamic task allocation: a practical game-theoretic approach. In: Proceedings of AAMAS, pp. 915–922 (2009)
7. Edalat, N., Xiao, W., Roy, N., Das, S.K., Motani, M.: Combinatorial auction-based task allocation in multi-application wireless sensor networks. In: The 9th International Conference on Embedded and Ubiquitous Computing (EUC), pp. 174–181 (2011)
8. Foster, I., Kesselman, C., Tuecke, S.: The anatomy of the grid: enabling scalable virtual organizations. Int. J. High Perform. Comput. Appl. **15**(3), 200–222 (2001)
9. Kraus, S., Shehory, O., Taase, G.: Coalition formation with uncertain heterogeneous information. In: Proceedings of AAMAS, pp. 1–8 (2003)

10. Macarthur, K., Stranders, R., Ramchurn, S., Jennings, N.: A distributed anytime algorithm for dynamic task allocation in multi-agent systems. In: Proceedings of AAAI, pp. 356–362 (2011)
11. McAfee, R.P., McMillan, J.: Auctions and bidding. J. Econ. Lit. **25**(2), 699–738 (1987)
12. Ramchurn, S., Polukarov, M., Farinelli, A., Truong, C., Jennings, N.: Coalition formation with spatial and temporal constraints. In: Proceedings of AAMAS, pp. 1181–1188 (2010)
13. Ramchurn, S.D., Farinelli, A., Macarthur, K.S., Jennings, N.R.: Decentralized coordination in robocup rescue. Comput. J. **53**(9), 1447–1461 (2010)
14. Regev, O., Nisan, N.: The popcorn market. Online markets for computational resources. Decis. Support Syst. **28**(1), 177–189 (2000)
15. Sen, S.: A comprehensive approach to trust management. In: Proceedings of AAMAS, pp. 797–800 (2013)
16. Shehory, O., Kraus, S.: Methods for task allocation via agent coalition formation. Artif. Intell. **101**(1–2), 165–200 (1998)
17. Wolski, R., Plank, J., Brevik, J., Bryan, T.: Analyzing market-based resource allocation strategies for the computational grid. Int. J. High Perform. Comput. Appl. **15**(3), 258–281 (2001)
18. Yu, B., Singh, M.P.: An evidential model of distributed reputation management. In: Proceedings of AAMAS, pp. 294–301 (2002)
19. Zheng, X., Koenig, S.: Reaction functions for task allocation to cooperative agents. In: Proceedings of AAMAS, pp. 559–566 (2008)

A Performance Optimization Support Framework for GPU-Based Traffic Simulations with Negotiating Agents

Yoshihito Sano, Yoshiaki Kadono and Naoki Fukuta

Abstract To realize a simulation which can handle hundreds of thousands of negotiating agents keeping their detailed behaviors, massive amount of computational power is required. Also having good programmability of agents' codes to realize complex behaviors is essential to realize it. On deploying such negotiating agents on an agent simulation, it is important to be able to handle detailed behaviors of them, as well as having a large scale simulation to cover important phenomenon that should be observed. There are strong demands to utilize GPU-based computing resources to handle large-scale but very detailed simulations. However, it is not easy task for developers to configure the sufficient parameters to be set on its compilation or execution time, analyzing their performance characteristics on various execution settings. In this paper, we present a framework to assist the coding process of negotiating agents on a traffic simulation, as well as its parameter tuning process on GPU-based programming for simulation developers to utilize GPGPU-based many parallel cores in their simulation programs efficiently. We show how our implemented prototype framework helps simulation developers optimize various parameters and coding-level optimizations to be run on various hardware and software settings.

1 Introduction

Multi-agent simulations have been applied to various fields, including traffic analyses [2], crowd simulations [27], planning an evaluation of evacuation plans on an airport [21], etc. On deploying a good agent simulation, it is important to be able to

Y. Sano (✉) · Y. Kadono · N. Fukuta
Graduate School of Informatics, Shizuoka University, Johoku,
Hmamamatsu 432-8011, Japan
e-mail: gs13017@s.inf.shizuoka.ac.jp

Y. Kadono
e-mail: gs14014@s.inf.shizuoka.ac.jp

N. Fukuta
e-mail: fukuta@cs.inf.shizuoka.ac.jp

© Springer International Publishing Switzerland 2016
N. Fukuta et al. (eds.), *Recent Advances in Agent-based Complex
Automated Negotiation*, Studies in Computational Intelligence 638,
DOI 10.1007/978-3-319-30307-9_9

handle detailed interactions among agents in a simulation. For example, [21] showed that the proposed agent simulation for an evacuation scenario in an airport, could identify a case that some people may not be able to escape from a danger situation even that scenario was considered as a safe case for evacuation. To analyze how vehicles and people act in unusual situations such as disasters or events, or to investigate the case that the precise behaviors of people might affect greatly in the overall behaviors, agents need to handle appropriate communications among each other, as well as dynamically responding to their environmental changes. Cooperation among agents have to be applied to not only such unusual situations but also a situation that each agent should dynamically respond to environmental changes. In such cases, coordination mechanism and replanning of agents should be done within fine grained cycles in a simulation [4]. For example, in [18], a negotiation-based technique has been applied to avoid heavy traffic congestions.

Another promising issue is to realize a large-scale multi-agent simulation while keeping such detailed interactions among agents in there. In actual situations, when realizing a traffic simulation that covers just one city, it is often necessary to handle millions of vehicles in the simulation to reproduce the actual phenomenon happening there, including traffic inbound to and outbound from the city. For example, a multi-agent simulation was proposed which considers what kind of influences were obtained when a city prepares lanes that electric vehicles are only allowed to run on them [9]. In [9], the simulation has been performed using approximately 3 million of agents.

One of the important key issues to be investigated is to make a multi-agent simulation large scale and increases the efficiency of execution [14]. However, to realize a simulation which is handling hundreds of thousands of negotiating agents to analyze behaviors of them on a very large-scale complex environment (e.g., an airport, a crowded train station or a whole megacity) with credible behaviors, it requires massive amount of computational resources.

Our primary goal is to make it easy to develop and deploy highly-scalable multi-agent simulations by utilizing codes that can be executed in massively parallel to allow those agents to cooperate each other and respond to their dynamic environmental changes on GPUs (Graphics Processing Units) or other modern computation infrastructures that were not recognized as a resource to be used for such purposes. General agent frameworks allow developers to prepare and reuse programs to realize complex behaviors for the agents. For example, JADE [19] provides a framework to implement FIPA-ACL-compliant interactions among agents. GENIUS [13] provides a platform to develop and analyze negotiating agents. Although GENIUS itself did not provide a platform to handle a large number of concurrently-negotiating agents, those agent programs can be applied in such situations by slightly extending them by the help of [23],[1] etc. However, it is not easy for those generic agent frameworks or platforms to utilize GPUs or such computation resources since often those agent codes could not directly be run on such resources. Furthermore, they require additional tasks to configure sufficient compilation and execution parameters, obtain

---

[1]On it, at least 4000 agents can be deployed, although it was not clearly mentioned in [23].

optimal coding practices for the codes, and analyze their performance characteristics for various execution conditions.

In this paper, we present a framework to realize better coding and parameter tuning experiences in GPU-based programming for negotiating agents to utilize GPGPU-based many parallel cores in their traffic simulation programs efficiently. We show how our prototype platform can be implemented and how our framework is helpful to optimize the parameters and coding practices on various hardware and software settings for realizing large-scale cooperative agents simulations.

The rest of the pager is organized as follows. Section 2 describes related work about large-scale simulations and GPU-related techniques. In Sect. 3, we show the overview of our proposed framework. Section 4 shows the implementation of the proposed framework, and discusses the merits and difficulties on the optimization. We conclude and discuss the future work in Sect. 5.

2 Related Work

In the viewpoint of tradeoffs among the complexly and granularity of the simulation, often some simplifications should be done. For example, the behavior of each car agent can be simplified to cover the phenomenon of daily traffic to handle a whole the city in the simulation [2]. Extending the scale of a multi-agent simulation is presented [27]. The method proposed in [27] limits the possible search space which an agent is going to move within a linked nodes instead of considering the whole 3-dimension spaces. In [27], the authors showed that it can dramatically reduce the calculation of collisions among agents. The method can help realize a large-scale multi-agent simulation by effectively simplifying the details in an agent simulation. There are several works based on the idea of changing degrees of details in behavior of agents to reflect what should be examined for the user's needs to handle a large-scale multi agent simulation (e.g., [15]). However, to decide sufficient degrees of details in a simulation, often such degrees were decided not based on its demand but rather from its actual computational performance. Therefore, it is still important to raise up the base computational performance to be applied to such a simulation.

A number of approaches have been proposed to realize efficient processing of a large-scale traffic simulation by massively parallel computers. In [26], as a base for performing an agent simulation, agent server IBM Zonal Agent-based Simulation Environment (ZASE) has been developed which can efficiently run thread-level parallel programs. ZASE combines two or more agent servers accelerated by thread-level parallel executions, as well as decomposing the agent simulation into multiple processes that can be run on massively parallel computers to realize a large-scale agent simulation. However, the approach can normally be applied to SMP-based scalar processor computer clusters.

To improve scalability of a simulation, the use of cloud computing infrastructures and frameworks, (e.g., Hadoop, etc.) could also be effective. Since a huge amount of communications are necessary to synchronize data among distributed processes, it is crucial to keep their network's latency low and give enough bandwidth for them to keep scalability. In addition, the developers who want to perform a simulation cannot always prepare massive amount of computers with such low-latency networks. In this paper, we would initially focus on how a large-scale simulation could be done in a single or a small number of computers each of which has many computation cores.

GPU (Graphics Processing Unit) has been widely used to realize such high-performance computing scenarios, especially on graphics-related operations. In order to effectively utilize the rich computation resources provided by GPU on non-graphic operations, GPGPU (General Purpose GPU)-based programming models have been proposed.

When an agent-based simulation would be run on a GPGPU-based computing infrastructure, the codes for their internal processing have to be developed having special coding techniques and detailed parameter tunings for each specific runtime environment. In this paper, we focus on a framework which supports those coding, verification, and parameter tuning processes effectively.

Although our work initially targets a road traffic simulation, it could be extended to a generic simulation on a given network represented as a graph. To reproduce phenomenon caused by traffics of cars, humans, etc. on a simulation, their most common unit could be the movements of agents in a graph. In typical traffic simulations, an origin and a destination for each agent are given, and them the simulation engine reproduces each agents' moves toward the destination. To reproduce such moves in a simulation, a kind of graph search algorithms should be used to find out each agents' itineraries and paths to the destination.

A number of techniques have been investigated to find the shortest path of a large-scale graph using GPGPU [3, 25], and it has been reported that GPGPU can perform well to compute the shortest paths compared with CPU.

Although agents may perform replanning in order to respond to the dynamic environmental changes, such graph algorithms would be used so frequently. However, even when the shortest path search in a graph could be calculated on a GPGPU, we may need much intelligent and complex search algorithms rather than for a shortest paths in an agent simulation that aims at reproducing more realistic traffic behaviors, etc. In an agent simulation, an algorithm to be applied may be varied in their aims and their corresponding environments. In this paper, our framework enables the simulation code developers to develop and run a complicated planning and other processings using GPGPUs.

In the work [17], it has been shown that even when the processing has been done less than 256 travel agents, their processing times are mostly same in each experiment setting on parallel execution scenario in a GPU. From this result, it has been confirmed that the planning could be performed in parallel by multiple GPU cores. In addition, it has also been confirmed that the RTA* [12] and the LRTA* [8] have been performed similarly on their scalability performance. However, there is little work that consider inter-agent communications among them. In [4], it is shown that cooperative behaviors among agents may reduce traffic jams. It is still an open issue that which kind of cooperation could be applicable to what kind of problems. Here, as our initial step, we consider to use negotiation techniques among agents.

In [13], a platform is presented to develop and analyze bilateral negotiation strategies. Also in [22, 23], a framework is presented to extend such negotiations to be applied to concurrent bilateral negotiation scenarios. In this paper, we try to handle such negotiating agents effectively in traffic simulations.

Performance characteristics of GPUs are varied and they are deeply depend on their implementation, execution conditions, and executed codes. There could be some optimal parameters for every combination of algorithms and GPU types. In order to utilize computation power of GPU, it is necessary to appropriately optimize settings and codes to be processed on the GPU. For example, concerning CPU-GPU communication, there is a research which aims at improving the execution speed by optimizing communication of CPU and GPU [1]. In [11], the research presented an approach that can automatically optimize task partitionings for different problem sizes and different heterogeneous architectures, as well as optimizing heterogeneous distributed computing environment itself [5]. Furthermore, the approach in presented [6] showed a code generation scheme for stencil computations on GPU Architectures. Also a way to optimize control flows for better GPU utilization [7], and adaptive runtime optimization technique [24] have been proposed. Notice that, some optimization options may affect their computation accuracy to realize faster executions. This might cause some serious effects on its computation. Therefore, it is also important to verify how such optimization options will affect to a whole simulation by comparing visualized results on a realistic simulation setting.

3   Proposed Framework

There are several GPGPU-based computing platforms (e.g., CUDA, ATI Stream, OpenCL, etc.). In this paper, we initially focus on the use of OpenCL because of its easiness to learn and wider support of hardware platforms.

When we consider about building an agent simulation, the person who wants to develop a simulation is not necessarily a specialist in GPGPU-based programming. In this paper, we create a framework which can easily build and analyze a certain scale of agent simulations empowered by GPGPU's which also allows the developers to easily analyze and tune its execution speed for optimal executions on a certain hardware by presenting a kind of instant testing environment.

When agents should respond to dynamic environmental changes in the simulation, they should perform *re-planning*, e.g., a kind of behavior which re-determine the route to the destination in a car traffic simulation. Therefore, such a simulation consumes lots of computation resources.

GPU is good at performing SIMD-computation, which applies a single instruction to multiple data, as well as running such SIMD computing threads in parallel. On GPU processing, a core program of such parallel processing is called kernel program. Therefore, the code which performs the same instructions to multiple data in the replanning process should also be described as a kernel program. In addition, we should also consider the case that each path planning algorithm applied to each agent may differ in order to express each agent's behavior. Therefore, we focus on improvement in the speed of the whole simulation including the planning for every agent can do parallel processing by using GPU rather than presenting a fast search and planning algorithm that could run faster on GPUs.

In the OpenCL programming model, there are two basic types of parallel processing; data parallel processing and task parallel processing. Data parallel processing performs single instruction on each processor to multiple data. The efficiency of computation on data parallel approach is often faster by applying similar computations to multiple data at once. In our case, this 'data parallel' approach can be applied to path planning and negotiation handling for every agent.

Figure 1 shows the structure of proposed framework. Our framework has five modules: Simulation Module, Real-time Rendering Module, Benchmarking Module, User Interface Module, and Parameter Tuning Module. Simulation Module runs the codes on light-weight traffic simulation with developer's kernel code and Real-time Rendering Module shows the real-time situation of the simulation. Benchmarking Module is a module to test and evaluate performance and scalability of kernel functions which will run on GPGPUs. The User Interface Module is a frontend to operate various functions implemented in our framework. Parameter Tuning Module is a module to semi-automatically adjust to optimal parameter set for each GPU and their simulation settings.

The developers using our framework can use OpenCL programming model within a C program, and various kinds of computations for the agents, e.g., path planning can be described. The data of a road network, the number of agents, etc. can be received as arguments of the specific kernel functions, and the developer can describe

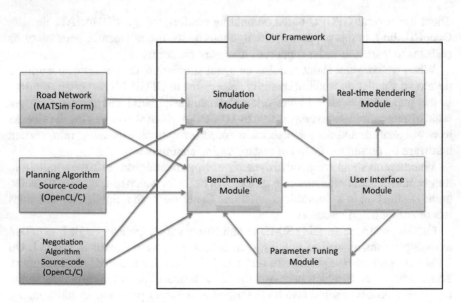

Fig. 1 The structure of proposed framework

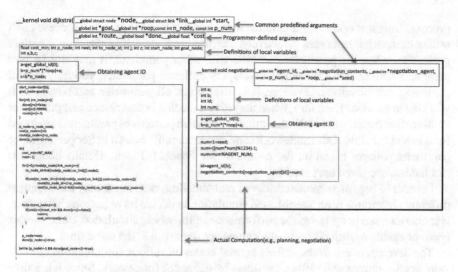

Fig. 2 An Example Code using the Framework

various programs that will access those data. When the coding of kernel programs are completed, the developer then register them to the runtime platform by simply specifying their function names. By doing so, developers can perform a test-simulation with the newly developed path planning, negotiations, etc.

Figure 2 shows an example code of an agent using the proposed framework. Programmers can use common arguments in order to receive the data to be processed (e.g., map data etc.) from the testing environment. Also, programmers can define the user-defined arguments, such as work domain used in the planning process, negotiation process, etc. The user-defined arguments can be used by describing the types, memory sizes to be used, and the passed information to the framework. Then programmers can distinguish each agents by using agent ID assigned by the execution platform. Agent ID is to distinguish an agent which processes with a core with GPU. After completing these descriptions, programmers can describe the main part of planning algorithms, negotiation process, etc.

Agents can communicate with other agents by using negotiation code. The developers write negotiation code about message for sending other agents. Each Agent has a own memory place. By writing information to agent's memory place, an agent can send message to another agents.

When the developers want to utilize GPGPU-based many parallel cores in their kernel functions, there are some parameters to be optimized [10]. Therefore, when the developers evaluate scalability of kernel functions by using our framework, they can adjust the maximum parallel threads, memory arrangement on a program, etc. The developers can see the test results in the parameters that they configured by using the framework. Even when optimizing one parameter, there are many values which can be set as a parameter, and the influence may differ for every environment. In addition, the parameter that is optimized for a certain process may not be effective in other process. Unless it optimizes a parameter for every target process, the process can't utilize computing resources effectively. Furthermore, even if optimizing a certain parameter, when the parameter combine with other parameter, it may not be best parameter setting. Thus when the developers optimize parameter, you may test all parameter combinations. However, if it tries to test all parameter combinations, It will take huge time. Therefore, when they give a file that is describes some parameter combination patterns to our framework, the specified patterns of parameters are tuned for a kernel function. Our framework semi-automatically seeks a better performance parameter patterns based on the results of the conducted tests. Details about this mechanism are given next section.

In order to improve reproducibility of real situation, each agent may have applied different algorithms in an agent-based simulation to decide its behaviors. Therefore, it is also an issue to optimize the performance of the whole simulation when various types of agents having different algorithms are deployed at the same time.

The developers can evaluate their agents' codes on various combinations of hardware configurations and software settings by using our framework. Since it is a time consuming task for developers to evaluate and set various parameters manually for every environments, we extended our framework to be deployed on a networked environment and run necessary measurements that should be done on different computers, etc. Setting scenarios of tests to the execution environments could be handled by the management console of our framework. Each execution environment conducts the evaluation and does the auto-tuning process according to the scenario. The results are finally summarized and the developers can get the optimal parameters to be used for the environment.

4 Implementation

4.1 Overview

We implemented a runtime platform based on the framework we proposed in the previous section. Figure 3 shows the overview of the runtime platform that we implemented. It can perform a simple road traffic simulation, and perform an agent's internal processing (e.g., dynamic planning, sending and receiving offers, etc.) using OpenCL-based codings. We implemented four major path planning algorithms, Dijkstra, A*, RTA* [12], and LRTA* [8] as a sample code set for further investigation of the users of our framework.

We implement two semi-automated performance tuning mechanisms: full-measuring mode and approximate optimization mode. In the full-measuring mode, the performance has been measured on each execution and optimization parameter settings, and the specified all possible combinations of such parameters are examined. To efficiently examine and evaluate performances on a limited combinations of parameters, the problem could be modeled as a variant of multi-armed

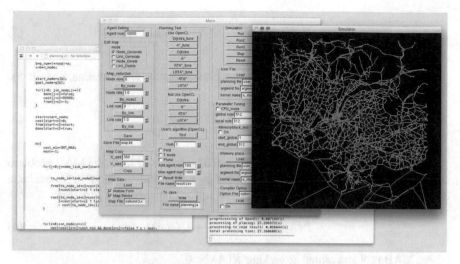

Fig. 3 The overview of execution environment

bandits [16] and a kind of algorithm to solve such a problem (e.g., a modified version of an algorithm appeared in [20]) could be applied. To realize such an efficient parameter optimization, we applied the following approach.

4.2 The Budget-Limited Multi-armed Bandits Model

The Multi-Armed Bandits (MAB) Problem defined in the Multi-Armed Bandits model [16] assumes that there is a slot machine with K arms and each arm has its own revenue distribution. In the situation, the problem seeks how an algorithm can choose the best arm to obtain maximal revenues as well as the cost for estimating the revenue distributions of arms. The Budget-Limited Multi-Armed Bandits (BLMAB) model [20] extends the MAB model to handle limited budgets and costs for pulling arms to estimate revenue distributions. Here, each arm i has its own cost c_i to pull the arm, and the payment will be withdrawn from the total budget B. Here, let A an algorithm to choose arms, $N_i^A(B)$ be the number of pulling arm i, the constraint for BLMAB model is defined as follows:

$$P\left(\sum_i^K N_i^A(B)c_i \leq B\right) = 1 \tag{1}$$

Let μ_i be an average revenu to pull the arm i, $G(A)$ be the total revenue by algorithm A, the expected value for $G(A)$ is defined as follows:

$$\mathbb{E}[G(A)] = \sum_{i}^{K} \mathbb{E}[N_i^A(B)]\mu_i \tag{2}$$

Then, we can define the solution A^* that maximizes the total revenue, as follows:

$$A^* = \underset{A}{\operatorname{argmax}} \sum_{i}^{K} \mathbb{E}[N_i^A(B)]\mu_i \tag{3}$$

Here, although the A^* need to know the set of values μ_i, i in K, the MAB problem assumes that an algorithm A does not know μ_i before starting its trial. To evaluate an algorithm A we can use A^* by giving the following regression function:

$$R(A) = \mathbb{E}[G(A^*)] - \mathbb{E}[G(A)] \tag{4}$$

An MAB problem aimed at seeking $R(A)$ be 0.

4.3 Extending BLMAB Model to Handle Multi-dimension Costs

The Extended BLMAB model for the optimization parameter search problem is an extended BLMAB model, which introduces a two-dimension cost constraint, i.e., the cost for examining performance improvements and the cost for using the computing device to examine the performance improvement, e.g., the cost paid for its power consumption. In our extended BLMAB model, the player can choose an arbitrary number of arms from the total K arms to pull in each step. Each arm i has its cost c_i to obtain the revenue based on an unknown distribution μ_i. The player has budget B so that the cost to be payed should be in the budget, and also the maximum steps is limited to be $T \in N$. The objective of the player is maximizing its revenue by choosing arm i with the limited budget B within the limited T steps. Therefore, we extend the original BLMAB constraints to the following:

$$P\left(\sum_{i}^{K} N_i^A(B)c_i \leq B \cap t \leq T\right) = 1 \tag{5}$$

4.4 The Basic Idea

On the optimization parameter search problem, we can assign each optimization parameter as each arm in the extended BLMAB model. Here, often each parameter has a limited number of variations in its effective performance tuning examination.

Therefore, the extended BLMAB model should satisfy that, let Q be the number of variations in its performance tuning examination, an arm i should satisfy $N_i^A(B) \leq Q$.

While the main objective of MAB is to maximize the total revenue, the main objective in optimization parameter search problem is to find the best parameters to be turned. Therefore, let $i(A)$ be the best arm obtained from A, $i(A*)$ be the theoretically best arm, we re-define the regression function for our extended BLMAB model as follows:

$$R(A) = i(A^*) - i(A) \qquad (6)$$

Note that, as we mentioned, the goal is to discover the best parameter set that generates the best performance, rather than total revenues obtained by performance tuning examination.

4.5 Applying ε-Greedy Algorithm

The ε-greedy algorithm is a hill-climbing random search algorithm, which follows greedy approach in the probability $(1 - \varepsilon)$, but does a random search in the probability ε.

The Algorithm 1 is an ε-greedy-based optimization parameter search algorithm based on our extended BLMAB model.

Algorithm 1 Extended Algorithm based on ε-greedy

Require: $0 \leq \varepsilon \leq 1$
1: $t = 1, B_t = B, T_t = T$
2: **while** $B_t > min_i c_i and t \leq T$ **do**
3:    let $s(B_t, T_t)$ be the number of concurrent executions
4:    **for** $j = 1$ to $s(B_t)$ **do**
5:      **if** $(1 - \varepsilon) > random \in [0, 1]$ **then**
6:       add the best parameter name evaluated by $H_i(t)$ to $\{i(t)\}$
7:      **else**
8:       add a randomly chosen applicable parameter name to $\{i(t)\}$
9:      **end if**
10:   **end for**
11:   run the parameters in the set $\{i(t)\}$
12:   update B_t and t
13: **end while**
14: decide best arm;

Here, for each time $t \in T$, B_t be the rest of budget in t, $i(t)$ be the set of performance tuning examination to be used in t. The line 2 in algorithm1 shows the continuation condition of the algorithm. From lines 6 to 8 in the algorithm, we use a function to know which arms can be chosen in the step t, satisfying the constraint that when the performance tuning examination for arm i is added it does not exceeds its budget constraint, as well as satisfying $N_i^A(B) \leq Q$.

We define the evaluation function $H_i(t)$ as follows:

$$H_i(t) = \frac{\hat{\mu}_{i,n_{i,t}}}{c_i} \tag{7}$$

In the above equation, $\hat{\mu}$ denotes the average revenue in time t to pull the arm i by executing its performance tuning examination.

4.6 Experiments

We conducted an evaluation for its parallel processing performance on the negotiations as well as path planning algorithms to validate the potential scalability improvements obtained by the proposed optimization support mechanisms. We examined the performance on the following scenario. On each planning problem, each agent receives an origin and a destination randomly. On each negotiation problem, each agent receives its negotiation partner and do its negotiation task for 100 cycles, i.e., sending and receiving alternatives 100 times. On *planning only* scenario, the above planning problem solving task is repeated 10 times. On *negotiation and planning* scenario, first it runs the 100 cycles of negotiation and then run the above planning problem solving task, then the scenario repeats the two tasks 10 times.

We measured the processing time for the whole processing where all agents completed their negotiation cycles and retrieved the route on various conditions in the scale, parameters, and parallelizing methods that can be specified on OpenCL.

We used a MacBook Pro(os: OS X 10.8, cpu: 2.4 GHz Intel Core 2 Duo, gpu: NVIDIA GeForce 320M, memory: 8GB 1067 MHz DDR3), a Mac Pro(os: OS X 10.8, cpu: 3.0Ghz Quad Core Xeon, gpu: Geforce 8800GT, memory: 32GB 800 MHz DDR2) and, a MacBook Pro(os: OS X 10.9 cpu: 2.6 GHz Intel Core i7, gpu: GeForce GT 750M, memory: 16GB 1600 MHz DDR3) as the experiment execution environment.

Table 1 shows a comparison of execution performance with or without memory-based optimizations in the code. We conducted the comparison on 10,000 agents with the map which has 384 nodes. Here, we can see when on the GeForce 320M and

Table 1 Comparison on performance with memory-assignment optimization

	GeForce 750M (ms)	GeForce 8800GT (ms)	GeForce 320M (ms)
dijkstra (optimized)	4623.52	7517.88	15662.69
dijkstra (not optimized)	6562.31	8726.93	30952.75
A*(optimized)	393.68	838.71	2901.36
A*(not optimized)	561.88	873.60	4169.71

Table 2 Performance optimization on planning and negotiation task: memory block divisions optimization

	NODE:384 (ms)	NODE:768 (ms)	NODE:1536 (ms)
A* with 1000 step negotiation (optimized)	7463.44	12859.16	23046.58
A* without negotiation (optimized)	3302.47	8782.62	19071.38
A* with 1000 step negotiation (not optimized)	9600.95	18507.46	36503.29
A* without negotiation (not optimized)	5305.03	14233.73	31889.85
A* with 1000 step negotiation (worst)	12954.06	24234.65	49989.25
A* without negotiation (worst)	8256.67	19962.48	44717.90

Table 3 Performance on negotiation-only task with different optimization parameters

	NODE:384 (ms)	NODE:768 (ms)	NODE:1536 (ms)
Negotiation-only (*optimized* for planning)	4772.98	4753.69	4879.81
Negotiation-only (*not optimized* for planning)	5123.60	5141.15	5110.72
Negotiation-only (*worst* for planning)	4777.86	4338.66	4442.57

GeForce 750M, the optimization done by our framework works very well. However, when the execution was done on GeForce 8800 GT, its effectiveness is very low. This shows how the optimization of code affects its performance and how it depends on its execution environment. For this reason, we prepared an execution and benchmarking functionality on our platform.

Table 2 shows the optimization performance comparison on the scenarios with or without negotiations. Here, *optimized* means the case when the optimal parameters obtained by the framework have been used for its execution, and *worst* means the case when the worst parameters obtained by the framework have been used. The case *not optimized* means that it just uses a parameter set that one of the authors usually used as a heuristically *best* parameter. We can see that the performance has been well improved when the scale of planning tasks are enlarged. Here, since we assume all agents will do the planning task all cycles, relatively large computation time has been spent for its planning tasks when the node size is large. This could be changed when not all agents will do such planning tasks, and this often happens when only a limited number of agents have to change their routes based on the negotiation.

Table 3 shows the performance of negotiation-only task when the same parameter sets used in Table 2 are applied. Since the parameter sets are only optimal/not-optimal/worst on both negotiations and planning tasks have been done, those parameters will not produce optimal, etc. in this case. Note that, when the *worst-case* parameter set has taken, the negotiation performance is almost same as its *optimized-case*, while the *not-optimized-case* produces approximately 10 % worse performance than the *worst-case*. This means that the heuristically good parameters were very different for its tasks and we may need a very difficult parameter optimization when the tasks are mixed. Our framework could be helpful to optimize such mixed scenario cases.

To analyze the detailed behaviors of computations in the GPUs, we might need to have a help of GPU-depend detailed profilers to obtain the reason of bottlenecks. To make a strong coupling of such GPU-depend profilers with our framework is future work.

5 Conclusion

In this paper, we presented a framework to assist the coding and parameter tuning process of GPU-based programming for negotiating agent-based simulations that utilizes GPGPU-based many parallel cores in their simulation programs efficiently. We presented a preliminary case study on a traffic simulation with concurrent bilateral negotiating agents which can negotiate each other and do frequent route replannings. We initially prepared an OpenCL-based implementation on our runtime platform to assist the code and parameter tuning. By using the proposed framework, when the agent's actions are coded using OpenCL, our framework could reduce the load in analyzing characteristics of each GPU types, parameters, etc.

Future work includes an evaluation of effectiveness of our proposed framework on various scalability improvement scenarios on a specific simulation problem. In addition, to make it easy to utilize two or more GPU(s) on our framework is another future work. There could be several approaches utilizing two or more GPU(s). For example, use two or more GPU(s) in one computer is a possible scenario, and also to run on two or more computers each of which has single GPU is another future work, and their combinations would be necessary. Since it is not easy to prepare all possible settings as actual execution environment, it could be helpful to collect some key characteristics from machines with some typical configurations and then make the framework predicts possible performances for a specific (or optimal) configuration of equipments.

References

1. AlSaber, N., Kulkarni, M.: Semcache: semantics-aware caching for efficient GPU offloading. In: Proceedings of the 27th ACM International Conference on Supercomputing (ICS '13), pp. 421–432, New York, NY, USA. ACM (2013)
2. Balmer, M., Meister, K., Rieser, M., Nagel, K., Axhausen, K.: Agent-based simulation of travel demand: structure and computational performance of matsim-t. In: 2nd TRB Conference on Innovations in Travel Modeling (2008)
3. Caggianese, G., Erra, U.: GPU accelerated multi-agent path planning based on grid space decomposition. In: Proceedings of the International Conference on Computational Science, pp. 1847–1856 (2012)

4. de la Hoz, E., Marsa-Maestre, I., Lopez-Carmona, M.A., Perez, P.: Extending matsim to allow the simulation of route coordination mechanisms. In: Proceedings of the 1st International Workshop on Multi-Agent Smart Computing (MASmart 2011), pp. 1–15 (2011)
5. Grasso, I., Pellegrini, S., Cosenza, B., Fahringer, T.: libwater: heterogeneous distributed computing made easy. In: Proceedings of the 27th ACM International Conference on Supercomputing (ICS '13), pp. 161–172, New York, NY, USA. ACM (2013)
6. Holewinski, J., Pouchet, L.-N., Sadayappan, P.: High-performance code generation for stencil computations on GPU architectures. In: Proceedings of the 26th ACM International Conference on Supercomputing, ICS '12, pp. 311–320, New York, NY, USA. ACM (2012)
7. Huo, X., Krishnamoorthy, S., Agrawal, G.: Efficient scheduling of recursive control flow on GPUs. In: Proceedings of the 27th ACM International Conference on Supercomputing (ICS '13), pp. 409–420, New York, NY, USA. ACM (2013)
8. Ishida, T., Shimbo, M.: Path learning by realtime search. Jpn. Soc. Artif. Intell. 11(3), 411–419 (1996). (In Japanese)
9. Kanamori, R., Morikawa, T., Ito, T.: Evaluation of special lanes as incentive policies for promoting electric vehicles. In: Proceedings of the 1st International Workshop on Multi-Agent Smart Computing (MASmart 2011), pp. 45–56 (2011)
10. Khronos OpenCL Working Group. The OpenCL Specification Version: 1.2 Revision: 19 (2012)
11. Kofler, K., Grasso, I., Cosenza, B., Fahringer, T.: An automatic input-sensitive approach for heterogeneous task partitioning. In: Proceedings of the 27th ACM International Conference on Supercomputing (ICS '13), pp. 149–160, New York, NY, USA. ACM (2013)
12. Korf, R.E.: Real-time heuristic search. Artif. Intell. 42(2–3), 189–211 (1990)
13. Lin, R., Kraus, S., Baarslag, T., Tykhonov, D., Hindriks, K., Jonker, C.M.: Genius: an integrated environment for supporting the design of generic automated negotiators. Comput. Intell. (2012)
14. Nakajima, Y., Yamane, S., Hattori, H.: Multi-model based simulation platform for urban traffic simulation. In: 13th International Conference on Principles and Practice of Multi-Agent Systems (PRIMA 2010), pp. 228–241 (2010)
15. Navarro, L., Corruble, V., Flacher, F., Zucker, J.-D.: A flexible approach to multi-level agent-based simulation with the mesoscopic representation. In: Proceedings of the International Conference on Autonomous Agents and Multiagent Systems (AAMAS 2013), pp. 159–166 (2013)
16. Robbins, H.: Some aspects of the sequential design of experiments. Bull. Am. Math. Soc. 58(5), 527–535 (1952)
17. Sano, Y., Fukuta, N.: A GPU-based framework for large-scale multi-agent traffic simulations. In: Proceedings of the 2nd IIAI International Conference on Advanced Applied Informatics (IIAI AAI2013) (2013)
18. Takahashi, J., Kanamori, R., Ito, T.: Evaluation of automated negotiation system for changing route assignment to acquire efficient traffic flow. In: Proceedings of the IEEE International Conference on Service Oriented Computing and Applications (SOCA2013), pp. 351–355 (2013)
19. Tilab. Java Agent Development Framework. http://jade.tilab.com
20. Tran-Thanh, L., Chapman, A.C., Rogers, A., Jennings, N.R.: Knapsack based optimal policies for budget-limited multi-armed bandits. In: AAAI (2012)
21. Tsai, J., Fridman, N., Bowring, E., Brown, M., Epstein, S., Kaminka, G., Marsella, S., Ogden, A., Rika, I., Sheel, A., Taylor, M.E., Wang, X., Zilka, A., Tambe, M.: Escapes—evacuation simulation with children, authorities, parents, emotions, and social comparison. In: Proceedings of the International Conference on Autonomous Agents and Multiagent Systems (AAMAS 2011), pp. 457–464 (2011)
22. Tsuruhashi, Y., Fukuta, N.: An analysis framework for meta strategies in simultaneous negotiations. In: Proceedings of 6th International Workshop on Agent-based Complex Automated Negotiations (ACAN2013) (2013)
23. Tsuruhashi, Y., Fukuta, N.: A framework for analyzing simultaneous negotiations. In: 16th International Conference on Principles and Practice of Multi-Agent Systems (PRIMA 2013) (2013)

24. Vasudevan, R., Vadhiyar, S.S., Kalé, L.V.: G-charm: an adaptive runtime system for message-driven parallel applications on hybrid systems. In: Proceedings of the 27th ACM International Conference on Supercomputing (ICS '13), pp. 349–358, New York, NY, USA. ACM (2013)
25. Vineet, V., Harish, P., Patidar, S., Narayanan, P.J.: Fast minimum spanning tree for large graphs on the GPU. In: Proceedings of the Conference on High Performance Graphics 2009, HPG '09, pp. 167–171, New York, NY, USA. ACM (2009)
26. Yamamoto, G., Tai, H., Mizuta, H.: A platform for massive agent-based simulation and its evaluation. In: Proceedings of the International Conference on Autonomous Agents and Multiagent Systems (AAMAS 2007), pp. 900–902 (2007)
27. Yamashita, T., Okada, T., Noda, I.: Implementation of simulation environment for control of huge-scale pedestrian. In: Joint Agent Workshop and Symposium (JAWS) (2012) (In Japanese)

RECON: A Robust Multi-agent Environment for Simulating COncurrent Negotiations

Bedour Alrayes, Özgür Kafalı and Kostas Stathis

Abstract RECON is an experimental simulation platform that supports the development of software agents interacting concurrently with other agents in negotiation domains. Unlike existing simulation toolkits that support only imperative negotiation strategies, RECON also supports declarative strategies, for applications where logic-based agents need to explain their negotiation decisions to a user. RECON is built on top of the GOLEM agent platform, specialized with a set of infrastructure agents that can manage an electronic market and extract statistics from the negotiations that take place. We evaluate the performance of RECON by showing how by increasing the number of agents in a simulation affects the agents' time to make an offer during negotiation.

Keywords Electronic markets · Automated negotiation · Simulation testbed

1 Introduction

Electronic markets (e-markets) are gaining increasing popularity among applications where components like software agents can act on behalf of buyers/sellers in order to buy/sell products over a distributed network [9]. In recent years, we have witnessed a huge number of successful e-markets on the Web, such as E-Bay, Amazon or Gumtree. E-markets of this kind have become popular because they support mechanisms such as advertising, auctioning and paying for goods online, thus providing an efficient and convenient way to perform commercial activities on the Web.

B. Alrayes (✉) · Ö. Kafalı · K. Stathis
Department of Computer Science, Royal Holloway, University of London,
Egham Hill, Egham TW20 0EX, UK
e-mail: b.m.alrayes@cs.rhul.ac.uk

Ö. Kafalı
e-mail: ozgur.kafali@cs.rhul.ac.uk

K. Stathis
e-mail: kostas.stathis@cs.rhul.ac.uk

© Springer International Publishing Switzerland 2016
N. Fukuta et al. (eds.), *Recent Advances in Agent-based Complex
Automated Negotiation*, Studies in Computational Intelligence 638,
DOI 10.1007/978-3-319-30307-9_10

157

One of the key capabilities that a software agent should possess when operating in an e-market is that of automated negotiation. Whether in a buyer or seller role, such capability allows the agent to bargain for goods by following a strategy, a plan of action designed to achieve a long-term objective.

One major issue in the design of a negotiation agent that participates in e-markets is how to evaluate the performance of its strategy. While some researchers test their agents theoretically [11], most agent developers use simulation platforms [3, 14, 18, 22, 26], especially for evaluating heuristic-based strategies. This gives rise to the need for a standardised simulation environment to provide fair and objective comparisons between negotiating agents. A successful negotiation platform should be (i) able to provide an open and dynamic environment for its participants, (ii) robust to the changes that occur in the market, (iii) reliable with the communication among agents, and (iv) scalable in terms of the number of agents it can support.

Recent negotiation literature suggests that most simulators are designed for specific domains with limited protocol and agent types [3, 14, 18, 22, 25, 26]. GENIUS [19] is a state-of-the-art negotiation platform that provides a competitive framework to test and compare bi-lateral negotiation strategies. Developers can implement their agent strategies using Java, following the provided negotiation protocol. However, the GENIUS code and its periodic extensions (e.g., see Williams et al. [26]) are not provided as open-source software. The system does not provide an open and dynamic e-market setting where buyers and sellers can enter/leave the system. The market options in GENIUS are very limited to describe a realistic setting. For example the system imposes fixed deadlines for the end of negotiation which is publicly available to both buyers and sellers. In addition, although GENIUS has been thoroughly tested using the ANAC competition series, it is unclear whether it can support markets with a large number of negotiating agents.

Motivated by the current limitations in GENIUS we develop RECON: a Robust multi-agent Environment for simulating COncurrent Negotiations. RECON supports the development of software agents (both buyers and sellers) negotiating concurrently with other agents over various issues. In contrast to most agent development platforms such as JADE and GENIUS which only support imperative (e.g. Java) agents, RECON supports both imperative (e.g. Java) and declarative (e.g. Prolog) strategies. Declarative strategies allow developers to specify strategies that can be transparent to a human user, in that the agent can explain why it has taken certain actions during a negotiation.

RECON is built using a simplified version of the GOLEM agent platform [4, 20], specialised with a set of infrastructure agents that manage an e-market and provide statistics over outcomes of simulation runs. We evaluate the performance and robustness of RECON using agents developed imperatively and declaratively. Our preliminary experiments with RECON report on e-market simulations with up to 200 agents negotiating with very reasonable offer-generation times. Moreover, we evaluate the system under different market settings by testing different simulation scenarios that experimenters might explore in practice.

The rest of the paper is organised as follows. Section 2 reviews the relevant background on GENIUS and our choice of GOLEM. Section 3 introduces RECON and describes its key features. Section 4 presents ours experiments to evaluate the performance of RECON with an increasing number of agents. Finally, Sect. 5 summarises the paper and outlines possible extensions for the future.

2 Background

The multi-agent literature is rich of negotiation models and strategies. However, there has not been much work in the area of simulation for negotiating agents. In this context, we may classify the relevant literature according to:

- *proprietary simulators*—which perform specific, closed (i.e., not open source) experiments that are developed by researchers to evaluate their own agent strategies, e.g. in bilateral negotiations [3, 21], concurrent bilateral negotiations [3, 18, 22], multilateral negotiation [23] and opponent models [8, 14];
- *public simulators*—which are generic, usually open-source simulators used to evaluate any agent strategy, as long as it is expressed in the system's specification language, e.g. the state-of-the-art GENIUS negotiation environment [19].

We discuss next GENIUS due to its wide use and acceptability as a publicly available simulation tool.

2.1 GENIUS

GENIUS is a negotiation environment that implements an open architecture for heterogeneous negotiating agents. It provides a testbed for negotiating agents that includes a set of negotiation problems for benchmarking agents, a library of negotiation strategies, and analytical tools to evaluate an agent's performance.

To verify the efficacy of GENIUS, the system was used by 65 students, who were required to design an automated agent each for different negotiation tasks. Their agents were evaluated and both quantitative and qualitative results were gathered. These results suggested that GENIUS helps and supports the design process of an automated negotiator (from the initial design, through the evaluation of the agent, and redesign and improvements) based on its performance.

To the best of our knowledge, GENIUS was used only for evaluating bilateral negotiation, especially agents in the automated negotiating agents competition (ANAC) [13]. In this context, Williams et al. extended GENIUS to provide support for concurrent negotiations [26]. However, this extension addressed a specific experimental setup and was not accessible publicly. Moreover, we are not aware of any work that evaluates the robustness and scalability of GENIUS using a large number of agents.

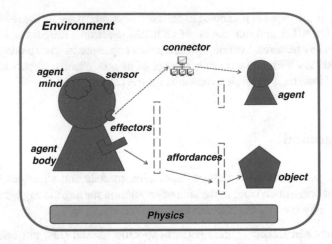

Fig. 1 GOLEM environment

Our initial intention was to extend GENIUS with extra functionality for concurrent negotiations as well as realistic e-market parameters. However, the above shortcomings of GENIUS discouraged us from trying to extend the system to serve our experimentation purposes. We thus decided to develop RECON on top of the GOLEM platform as it already supports agent deployment in different settings, including support for specifying declarative and/or imperative strategies negotiating agents.

2.2 The GOLEM Agent Platform

GOLEM[1] is a logic-based agent platform developed to represent agent environments that evolve over time [4, 5]. It provides a middleware that agent developers can use to build multi-agent systems using both Java and Prolog. The platform has been used to deploy agents in a number of practical applications, from ambient intelligence [7, 10] and service negotiation in grid computing [6], to diabetes monitoring and management [15]. As shown in Fig. 1, a GOLEM environment consists of three main components: containers, agents and objects.

- *Containers*—are logical entities representing a subset of the agent environment and mediating the interactions between agents and objects situated in it. Such interactions are governed by the container's physics, a component that prescribes how the state of the container changes as a results of agent actuators performing actions, including how actions and their effects are perceived by agent sensors. Using containers, an agent environment can be distributed over a network, allowing agents to perceive and interact with entities that are logically near, but physically

---

[1]http://golem.cs.rhul.ac.uk/.

distributed somewhere else. Containers may also contain sub-containers. A connector component attached to the container allows agents to communicate with each other.

- *Agents*—are cognitive entities that can sense the environment, reason about it and interact with other agents as well as objects. An agent is composed of an agent body and an agent mind. The body allows physical interaction with the environment via sensors and effectors, while the mind processes perceived knowledge to decide how the agent should act.
- *Objects*—are entities situated in the environment that react once agents act upon them.

In GOLEM the way entities make themselves present in the environment is described via the concept of *affordances*. Affordances are defined with ontologies and enable agents to perceive these entities in a systematic manner, without the need of a directory facilitator agent, see [4] for more details.

Agent development in GOLEM

Following previous work with platforms of logic-based agents [24], a GOLEM agent consists of two core components. The agent body, developed in Java, to allow interaction with the agent's surrounding environment through perceptions and actions. The agent mind, developed using logic-based techniques, to performs internal reasoning. In a basic reasoning cycle, when the body perceives a change in the environment through one of its sensors, it initiates a decision process to be performed by the mind. The mind decides what action to take based on the given internal state of the agent, and the body performs the selected action to finish the cycle.

Listing 1 presents a simple cycle theory [16] implemented in Prolog as an illustration of agent control. When a new percept is received via the body, the mind revises the internal state of the agent. Then, the mind decides what action to perform via the cycle step. First, it determines the current goal of the agent based on the new information received. Then, it selects an action to achieve that goal. For simplicity, the agent selects the first action whose conditions succeed in the state of the agent. The call to *once* uses the Prolog cut operator ! to achieve this. The ordering of the *select* rules describe which action must be preferred over others. The mind executes the action internally to revise the agent's goals and returns it to the body.

```
1    % perception
2    perceive(Percept, Time):-
3       revise(perceived(Percept), Time).

5    % cycle step
6    cycle_step(Action, Time):-
7       adopt(Goal, Time),
8       decide(Goal, Action, Time),
9       execute(Action, Time).

11   decide(Goal, Action, Time):-
12      once(select(Goal, Action, Time)).

14   % domain-dependent strategy
15   select(Goal, Action, Time):-
```

```
16        holds_at(Var1, Time,),...,holds_at(VarN, Time).
17        ...
18    execute(Action, Time):-
19        revise(attempted(Action), Time).
```

Listing 1 Simple agent mind in GOLEM

Note that GOLEM allows agent developers to make their own design choices when developing the agents. The developer can create the agent mind using a Prolog-driven approach, possibly supporting temporal reasoning [17]. More complex agent decision making can be developed in GOLEM, e.g. see [12]. In addition, the mind can be replaced by another programming language (like Java) for developers who do not wish logic-based strategies.

Agent communication

Containers are connected to each other via components called *connectors*. A connector is an abstract service that hides away the low-level details of message transportation. Connectors are registered to containers in order to enable agent communication, implemented by combining Prolog and Java.

When an agent needs to send a message to another agent, it simply produces a specific action described by the keyword *envelope*. Once the agent produces such an action, the connector that is registered with the agent's container picks it up, and transfers it to the recipient agent's connector. A variation of the send message is also used to send serialised objects between containers (e.g., to move an agent to a different physical machine).

2.3   GOLEMLite

GOLEMLite [20] is a Java library that maintains the key GOLEM concepts such as the container and agents with body, mind, sensors and effectors. However, in GOLEMLite there is no support for objects and applications can be deployed in one container with a stripped-down physics component that supports communicative actions between agents only (i.e. the system does not support physical actions).

3   RECON

RECON is a Robust multi-agent Environment for simulating COncurrent Negotiations. It is built on top of GOLEMLite and supports the development of negotiating agents for testing both buyer and seller strategies. These strategies can be developed either declaratively (implemented in Prolog) or imperatively (implemented in Java). Also, the system supports concurrent negotiations and dynamic markets, where agents can enter/leave the marketplace during negotiation. In addition, agents can have their private deadline during a negotiation, instead of a common public one, to reflect what happens in practice. Moreover, realistic applications require negotiations with both continuous and discrete issues, unlike supporting discrete and integer-valued issues only. Table 1 compares the functionality between RECON and GENIUS.

Table 1 Comparison of functionalities between RECON and GENIUS

Functionality	RECON	GENIUS
Open and dynamic	✓	✗
Support any protocol	✓	✗
Large number of agents	✓	✗
Support logic programming	✓	✗
Private deadlines	✓	✗
Continuous-valued issues	✓	✗
Human negotiators	✗	✓
Multiple issues	✗	✓
GUI	✗	✓

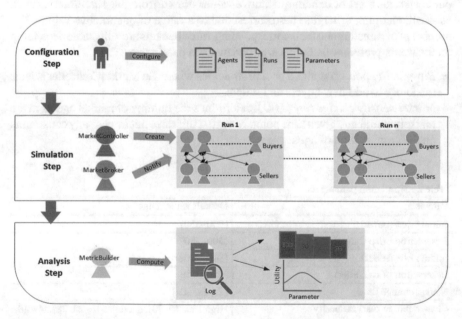

Fig. 2 The RECON architecture

A simulation in RECON assumes three steps (see Fig. 2):

1. In the configuration step, the user describes the simulation setting using the configuration tool.
2. In the simulation step, buyers and sellers negotiate with each other. The *Market-Controller* agent controls the execution while the *MarketBroker* agent acts as a yellow pages service for the buyers.
3. In the analysis step, simulation logs are inspected by the *MetricBuilder* agent to calculate related evaluation metrics.

Next, we describe each step in detail.

3.1 Configuration Step

In this step, the user defines the simulation parameters, the number of repeated runs in each parameter combination (RC), and the market agents. The simulation parameters, which describe the e-market setting, are summarised in Table 2. Each parameter has a set of default qualitative values drawn from the *ValueBuckets* such as small, medium, or large. These correspond to a range of quantitative values. The number of parameters and their corresponding value buckets are adjustable depending on the user's preferences. Let us now review each parameter:

- *Buyer/seller pool*: contains a pool of strategies where the MarketController selects randomly to add new buyers and sellers.
- *Market density change rate (MD)*: determines the number of market agents in the marketplace at any given time point. Note that this does not impose any constraints on the composition of agent types.

Table 2 Simulation parameters

Parameter	Default value range
Initial price (buyer)	[300–350]
Reservation price (buyer)	[500–550]
Initial price (seller)	[800–750]
Reservation price (seller)	[450–500]
Market change time	[2s, 5s, 10s]
Change rate in market density	High:[30, 40, 50], average:[18, 23, 28], low:[8, 10, 12]
Change rate in market ratio	High:[10:1, 1:1, 1:10], average:[5:1, 1:1, 1:5], low: [2:1, 1:1, 1:2]
Deadline	Short:[30s–90s], average:[91s–150s], long:[151s–210s]

- *Market ratio change rate (MR)*: determines the ratio of buyer agents to seller agents in the marketplace (i.e., demand/supply ratio). The value of market ratio and market density are used jointly to determine the number of the buyers and sellers in the marketplace based on the following formula:

$$NoBuyers = (MD * X)/(X + Y)$$

$$NoSellers = (MD * Y)/(X + Y)$$

For instance, if the value of Market Ratio is 2:1 and the market density is 30, then the number of buyers is $(30 * 2)/3 = 20$, and number of sellers is $(30 * 1)/3 = 10$.
- *Market change time (MC)*: represents the number of seconds at which the marketplace has to change. The change involves adding/removing buyers and sellers according to MD and MR. This parameter ensures an open environment, where participants are not necessarily known at design time.
- *Deadline (MDL)*: represents the deadline for all market agents. All buyers have the same deadline which is different from the sellers deadline. Moreover, the deadlines are kept private to the agents, so that buyers and sellers do not know each others' deadlines. The buyer's negotiation timer starts when it enters the marketplace. On the other hand, the seller has many negotiation timers: one per negotiation and another which starts when it receives the first offer from the buyer. Both buyers and sellers leave the marketplace when they reach their deadlines.
- *Initial price (IP)*: represents the ideal price for an agent to buy/sell an item. All buyers and sellers in each run will have the same initial price to provide fairness when comparing performance.
- *Reservation price (RP)*: represents the maximum (minimum) price the agent is willing to pay for (get from) an item. Similarly, all buyers and sellers have the same reservation price for fairness.

3.2 Simulation Step

The actual negotiation between the market agents takes place in this step. In general, the number of negotiation simulations RECON will conduct in each experiment depends on the parameters and RC from the configuration step. The following equation calculates the number of simulations:

$$NoSimulations = (\prod_{i=1}^{n} NoValueBuckets_i) * RC$$

where n is the number of parameters. For instance, to conduct a simulation for four parameters MD, MR, MC and MDL where each parameter has three values, and $RC = 100$, then the simulator will conduct 8100 (3 * 3 * 3 * 3 * 100) simulations.

Thus, this step will terminate when all the 8100 simulations finish. Likewise, each simulation ends when all of the selected buyers (i.e., the buyers that the user wants to measure the performances of) leave the marketplace, either because they have reached their goal or because they have reached the deadline. This step is managed by the following infrastructure agents (*MarketController* and *MarketBroker*), and played by *MarketAgents*.

3.2.1 Market Controller

The *MarketController* is an infrastructure agent that is accountable to manage all the simulations which will be run as part of the overall experiment. The Market-Controller runs the simulator for a fixed number of times defined by the user. At the end of each simulation, it logs the negotiation messages between MarketAgents. The MarketController has the following responsibilities:

- it creates the MarketBoker;
- it uses the configuration to initialise each simulation and creates all the required MarketAgents;
- during each simulation, it controls the number of MarketAgents in the marketplace by adding and removing them according to the market density and demand supply ratio;
- it asks all MarketAgents about their decisions at fixed time intervals;
- at the end of each simulation (i.e. when all Selected Buyers reach their goal or deadline), it stops and removes all agents, save negotiation history to a log file;
- when all simulations finish, the MarketController stops and removes the Market-Agents and the MarketBroker, and finally terminates itself.

3.2.2 Market Broker

The *MarketBroker* is an infrastructure agent created by the MarketController to match buyers with sellers. It acts as a yellow page service for the buyers where it notifies the existing buyers when a new seller enters the marketplace, and maintains a register of all MarketAgents in the marketplace. The MarketBroker maintains two registers internally: a list of buyers and a list of sellers. When a new seller enters the marketplace, the MarketBroker will send that seller's ID to every buyer that has registered interest in purchasing the product that the new seller is willing to sell.

3.2.3 Market Agents

The MarketAgents are agents constructed by the MarketController. Their goal is to allocate/offer a resource (i.e. item or service) from other MarketAgents by negotiating the resource issues (i.e. price). MarketAgents announce themselves when they enter the marketplace, specifying their class (buyer or seller) as well as the item they are demanding/offering. We followed the architecture of MarketAgents proposed in [2], as it shares our motivations.

Buyers are MarketAgents with the goal of maximising their utility from purchasing an item. They are liable for initiating negotiation with sellers notified by the MarketBroker. A typical buyer will remain in the marketplace until it reaches a successful negotiation, is told to leave by the MarketController or reach its deadline. Listings 2 and 3 in the Appendix describe part of the reasoning process of the buyer agent both in Java and Prolog.

Sellers are MarketAgents with the goal of maximising their utility from selling an item. Each seller negotiates with buyers that send them offers. A seller will leave the marketplace when the MarketController requests it or if its strategy dictates it must leave (e.g. because its inventory is empty). Sellers are responsible to terminate negotiation when their deadline is reached. Similarly, we can develop strategies for a seller as we have with buyers; their listings we omit due to lack of space.

3.3 Analysis Step

This step is controlled by the *MetricBuilder* agent which is responsible for analysing the log files generated from the simulation step.[2] The purpose of this step is to observe how market agents reason during negotiation, and to conclude why the agent is winning/losing. The MetricBuilder calculates a number of performance metrics defined by the user. The metrics are then plotted against various configuration parameters that describe the e-market setting. To reduce complexity, only the log of selected buyer agents will be parsed to provide results.

Currently, the following metrics are supported by RECON:

- *Number of successful runs* (SR_j): is the number of simulations per each parameter combination j where a negotiation ended in an *accept* action from the buyer.
- *Average utility per parameter combination* (AU_j): is the sum of the utility from each simulation divided by the number of runs (RC) in each parameter combination j. $AU_j \in [0, 1]$, where 0 indicates an unsuccessful negotiation and 1 denotes that buyer was able to acquire the item with the initial price.

$$U_i^j = (RP_i^j - AP_i^j)/(RP_i^j - IP_i^j)$$

$$AU_j = \sum_{i=1}^{RC} U_i^j / RC$$

where U_i^j is the utility for the jth parameter combination and ith repeated run, RP_i^j is the reservation price, IP_i^j is the initial price, AP_i^j is the accepted price.

---

[2]Listing 4 in the Appendix illustrates an example of a log file.

- *Average utility over successful runs* (AUS_j): is the sum of the utility from each simulation, divided by the number of successful runs. $AUS_j \in [0,1]$.

$$AUS_j = \sum_{i=1}^{RC} U_i^j / SR_j$$

4 Evaluation

To evaluate the performance of RECON and show that it provides a robust environment for negotiating agents, we conduct experiments to evaluate the cycle time of agents (e.g., the time it takes for the agent to generate an offer) in the following settings:

- The first set of experiments report average cycle time of agents throughout a simulation. The aim of these experiments is to show that agents are not affected by the execution of the simulation framework itself, i.e., they do not take longer time to decide on their actions towards the end of simulation.
- The second set of experiments report average cycle time of agents for increasing number of agents in the market. The aim of these experiments is to show that there is no exponential growth in agents' cycle times, i.e., having other agents in the environment has a linear affect on the time it takes for an agent to decide on an action.

RECON has already been used as a simulator to evaluate negotiating agents [1]. In those experiments, we have observed that the performance of RECON is steady and it can handle different agent types and market settings. Next, we provide the details of the experiments for the above settings and report the results.

4.1 Experimental Methodology

We have run the simulation for five diverse market densities (MD) up to 200 agents. For each density, 100 runs were performed. Therefore, a total of 500 runs were executed with each run having buyers and sellers negotiating concurrently. We have fixed the deadline to long $(MDL \in (151–210\,s))$, the market ratio change rate to high $(MR \in (10{:}1, 1{:}1, 1{:}10))$, and the market change time to quick $(MC \in (10\,s))$. The computer used for the experiment has the following specifications: Intel i5-3450S 2.80GHz processor with 8GB RAM running on Windows 7 Enterprise Service Pack 1 (64-bit Operating System).

Agent selection: For our experiments, we have developed an extended version of Faratin's strategies [3] that allow the sellers to concurrently negotiate with different buyers at the same time. For the buyers, we have selected three declarative and three imperative agents. We have recorded the average cycle times for each agent type.

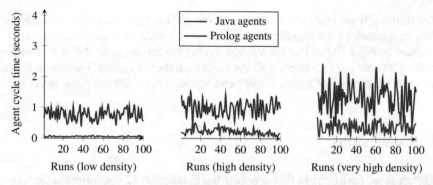

Fig. 3 Average cycle time for increasing number of runs

4.2 Results

The results for three of the five market densities are shown in Fig. 3. Low density implies that there are 8–12 agents in the marketplace. For high, there are 30–50 agents and for very high there are 100–130 agents. In each plot, the average cycle times for the three declarative and imperative buyers are consistent throughout all runs. Moreover, the changes in the market density does not affect the average cycle time. This conclusion is derived from the three plots where the average cycle time for the Java buyer stays within 0.03–0.7 s, and for the declarative buyer it stays within 0.3–2.4 s.

The fluctuations in the average cycle time is not caused by RECON, but it's rather an outcome of the buyers themselves. Since RECON supports asynchronous negotiation (i.e., where the buyer can negotiate at any time without waiting for all sellers reply), the buyers reply to different number of sellers. For instance, while the agent replies to one seller in one cycle, it might have to deal with three sellers in the next cycle

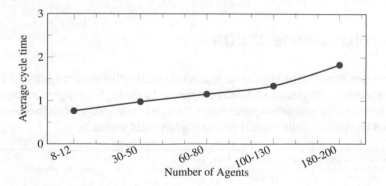

Fig. 4 Average cycle time for increasing number of agents

due to the sellers' reply times. Moreover, our buyers have heuristic strategies. So, they may check all the negotiation threads in one simulation, and not in the other.

Finally, Fig. 4 shows that the average cycle time for the three declarative buyers in the 100 runs grows linearly with increasing number of agents. We plan to further increase the number of agents in our experiments. This is left as future work.

5 Conclusions

RECON is an environment that supports the simulation of concurrent negotiations amongst multiple-agents in electronic markets. We have evaluated the performance of RECON with increasing number of agents in a market, and showed that its performance is stable, reliable and scalable under different simulation settings. Future work involves to:

- have a GUI which facilitate the use RECON;
- support multiple-issue negotiations by means of extending the market controller and market agents, handle discrete as well as continuous issues in order to compare our scenarios with GENIUS;
- include additional performance metrics in the analysis step;
- include human avatars in place for negotiating buyers and sellers (human avatars have already been developed in GOLEM to provide human interaction with agents);
- study experiments with larger number of agents;
- adopt a game theoretic approach as another option for evaluating the agent performance.

Acknowledgments We thank Ataul Munim for supporting us with developing the simulation environment. The first author wishes to acknowledge the support of a scholarship from the King Saud University.

Appendix: Agents in RECON

This section describes briefly the basic methods used by the market agents. Listings 2 and 3 presents examples of a Java buyer agent and a Prolog buyer agent, respectively. Note that the offer generation processes of the agents solely depend on their strategies and not on RECON. Seller agents can be implemented similarly.

```
1  public class SimpleBuyer extends AbstractBuyerAgent{
2    public SimpleBuyer(AgentBrain brain, AgentParameters params, String product){
3      super(brain, params, product);
4    }

6    @Override
7    protected List<Action> decideActionBasedOnOffer(NegotiationAction offer){
8      List<Action> actionsToPerform = new ArrayList<>();
```

```
9    double utilityOpponentOffer = getUtility(Double.parseDouble(offer.getValue()));
10   double counterOffer = generateNextOffer(offer.getDialogueId());
11   double utilityMyCounterOffer = getUtility(counterOffer);
12   boolean isOfferAcceptable = isOfferAcceptable(
13            utilityOpponentOffer,
14            utilityMyCounterOffer,
15            getNormalisedTime(getStartTime()));
16   if (isOfferAcceptable){
17     actionsToPerform.addAll(super.acceptOpponentOffer(offer));
18   }
19   else{
20     actionsToPerform.addAll(super.sendCounterOffer(offer, counterOffer));
21   }
22   return actionsToPerform;
23 }

25   @Override
26   protected List<Action> decideActionBasedOnAccept(NegotiationAction accept){
27     switch (getRandom().nextInt(NUM_RANDOM_ACTIONS)){
28     case ACCEPT:
29       return super.acceptOpponentOffer(accept);
30     case COMMIT:
31       return super.commitToMyLastBid(accept);
32     case EXIT:
33       return super.exitFromDialogue(accept.getDialogueId());
34     }
35     return new ArrayList<>();
36   }

38   @Override
39   protected double generateNextOffer(String dialogueId){
40     return getInitialPrice() + getRandom().nextInt(getReservationPrice() -
                getInitialPrice());
41   }

43   private boolean isOfferAcceptable(double utilityOpponentOffer, double
                utilityMyCounterOffer, double time){
44   }
```

Listing 2 Java buyer agent

In Listing 2, the Java buyer inherits the basic methods from AbstractBuyerAgent (line 1). The method *decideActionBasedOnOffer()* (line 7) returns an action based on the seller's offer. The method *getUtility()* (line 9) returns the utility of the seller offer. The method *generateNextOffer()* (line 10) calls the method in line 39 to generate the buyer next offer. This is part of the buyer strategy. The method *isOfferAcceptable()* (line 12) calls the method in line 43 to decide if the opponent's offer is acceptable by comparing the buyer offer with the opponent offer. This is part of the buyer strategy. The method *decideActionBasedOnAccept()* (line 26) returns an action based on the opponent accept action.

In Listing 3, *select(_, exit_all(Item), T)* (line 1) returns the action exit_all for all sellers that negotiate for item *Item* at time *T*. This decision will be taken if the buyer reaches its deadline. The predicate *select(Goal, offer(Opponent, Item, Offer),T)* (line 7) returns an offer (represented by the variable *Offer*) for the agent's goal (*Goal*), seller (*Opponent*) and item under negotiation (*Item*). This predicate calls a function to generate the counter offer at line 12 after checking the buyer's deadline. The predicate *calc_next_offer(buy(Id, Ao), Offer, T)* (line 15) generates the counteroffer at time *T* for negotiation dialogue (*Id*) based on: initial price (*Min*), reservation price (*Max*) and concession rate (*CA*). The predicate *concession(buy(Id, Ao), CA, T)* (line 22) calculates the concession rate. This part of the agent strategy.

```
1   select(_, exit_all(Item), T):-
2        deadline(Deadline),
3        ourdeadline(StartTime),
4        Td is (StartTime + Deadline),
5        T > Td.

7   select(Goal, offer(Opponent, Item, Offer),T):-
8        deadline(Deadline),
9        ourdeadline(StartTime),
10       Td is (StartTime + Deadline),
11       T < Td,
12       calc_next_offer(Goal, Offer, T).

14  % calculate an offer
15  calc_next_offer(buy(Id, Ao), Offer, T):-
16       % calculate concession rate
17       concession(buy(Id, Ao), CA, T),
18       holds_at(ip(Id, Min)=true, T),
19       holds_at(rp(Id, Max)=true,  T),
20       Offer is Min + (Max - Min) * CA.

22  concession(buy(Id, Ao), CA,  T):-
23       CA is 0.75.
```

Listing 3 Prolog buyer agent

```
TIMESTAMP:::FROM:::TO:::EVENT

1389744273493:::s_1:::buyer_1:::
    sell(0, ao),  offer(buyer_1, bananas, 768.3)
1389744273509:::buyer_2:::s_0:::
    buy(13, ao),  offer(s_0, bananas, 333.0)
1389744273509:::buyer_2:::s_1:::
    buy(1, ao),   offer(s_1, bananas, 332.0)
1389744273509:::buyer_2:::s_2:::
    buy(7, ao),   offer(s_2, bananas, 333.0)
1389744273509:::buyer_2:::s_3:::
    buy(3, ao),   offer(s_3, bananas, 335.0)
1389744273587:::s_3:::buyer_1:::
    sell(2, ao),  offer(buyer_1, bananas, 640.5)
1389744273603:::s_3:::buyer_2:::
    sell(3, ao),  offer(buyer_2, bananas, 640.5)
1389744273649:::s_0:::buyer_1:::
    sell(12, ao), offer(buyer_1, bananas, 495.0)
1389744273649:::s_0:::buyer_2:::
    sell(13, ao), offer(buyer_2, bananas, 776.5)
1389744273743:::buyer_1:::s_1:::
    buy(0, ao),   offer(s_1, bananas, 462.0)
1389744273743:::buyer_1:::s_3:::
    buy(2, ao),   offer(s_3, bananas, 430.0)
```

Listing 4 Narrative from a simulation run

References

1. Alrayes, B., Kafalı, Ö., Stathis, K.: CONAN: a heuristic strategy for COncurrent negotiating AgeNts (Extended abstract). In: Proceedings of the 2014 International Conference on Autonomous Agents and Multi-agent Systems (2014)
2. Alrayes, B., Stathis, K.: An agent architecture for concurrent bilateral negotiations. In: Impact of New Technologies in Decision-Making for Global Environments. Lecture Notes in Business Information Processing (LNBIP) (2014) (to appear)
3. An, B., Sim, K.M., Tang, L.G., Li, S.Q., Cheng, D.J.: Continuous-time negotiation mechanism for software agents. IEEE Trans. Syst., Man, Cybern., Part B: Cybern. 36(6), 1261–1272 (2006)
4. Bromuri, S., Stathis, K.: Situating Cognitive Agents in GOLEM, vol. 5049, pp. 115–134. Springer (2008)
5. Bromuri, S., Stathis, K.: Distributed agent environments in the ambient event calculus. In: Proceedings of the Third ACM International Conference on Distributed Event-Based Systems, DEBS '09, pp. 12:1–12:12, New York, NY, USA. ACM (2009)
6. Bromuri, S., Urovi, V., Morge, M., Stathis, K., Toni, F.: A Multi-agent system for service discovery, selection and negotiation. In: Sierra, C., Castelfranchi, C., Decker, K.S., Sichman, J.S. (eds.) 8th International Joint Conference on Autonomous Agents and Multiagent Systems (AAMAS 2009), Budapest, Hungary, May 10–15, 2009, vol. 2, pp. 1395–1396. IFAAMAS (2009)
7. Bromuri, S., Urovi, V., Stathis, K.: iCampus: a connected campus in the ambient event calculus. Int. J. Ambient Comput. Intell. (IJACI) 2(1), 59–65 (2010). doi:10.4018/jaci.2010010105
8. Brzostowski, J., Kowalczyk, R.: Predicting partner's behaviour in agent negotiation. In: Proceedings of the Fifth International Joint Conference on Autonomous Agents and Multiagent Systems, AAMAS '06, pp. 355–361, New York, NY, USA. ACM (2006)
9. Cohen, M., Stathis, K.: Strategic change stemming from e-commerce: implications of multi-agent systems in the supply chain. Strateg. Change 10, 139–149 (2001)
10. Dipsis, N., Stathis, K.: Ubiquitous agents for ambient ecologies. Pervasive Mobile Comput. 8(4), 562–574 (2012)
11. Fatima, S.S., Wooldridge, M., Jennings, N.R.: Multi-issue negotiation with deadlines. J. Artif. Int. Res. 27, 381–417 (2006)
12. Forth, J., Stathis, K., Toni, F.: Decision making with a KGP agent system. J. Decis. Syst. 15(2–3), 241–266 (2006)
13. Fujita, K., Ito, T., Baarslag, T., Hindriks, K., Jonker, C., Kraus, S., Lin, R.: The second automated negotiating agents competition (anac2011). In: Ito, T., Zhang, M., Robu, V., Matsuo, T. (eds.) Complex Automated Negotiations: Theories. Models, and Software Competitions, Studies in Computational Intelligence, vol. 435, pp. 183–197. Springer, Berlin (2013)
14. Hindriks, K., Tykhonov, D.: Opponent modelling in automated multi-issue negotiation using bayesian learning. In: Proceedings of the 7th International Joint Conference on Autonomous Agents and Multiagent Systems—Volume 1, AAMAS '08, pp. 331–338, Richland, SC (2008). International Foundation for Autonomous Agents and Multiagent Systems
15. Kafali, O., Bromuri, S., Aguilar-Pelaez, E., Sindlar, M., van der Weide, T., Schumacher, M., Rodriguez-Villegas, E., Stathis, K.: A smart e-health environment for diabetes management. J. Ambient Intell. Smart Env. 5(5), 479–502 (2013)
16. Kakas, A.C., Mancarella, P., Sadri, F., Stathis, K., Toni, F.: Declarative agent control. In: Leite, J.A., Torroni, P. (eds.) Computational Logic in Multi-Agent Systems, 5th International Workshop, (CLIMA V), Lecture Notes in Computer Science, vol. 3487, pp. 96–110. Springer (2005)
17. Kowalski, R., Sergot, M.: A logic-based calculus of events. New Gener. Comput. 4(1), 67–95 (1986)
18. Li, C., Giampapa, J.A., Sycara, K.: Bilateral negotiation decisions with uncertain dynamic outside options. IEEE Trans. Syst., Man, Cybern., Part C: Spec. Issue Game-theor. Anal. Stoch. Simul. Negot. Agents 36(1) (2006)

19. Lin, R., Kraus, S., Baarslag, T., Tykhonov, D., Hindriks, K., Jonker, C.M.: Genius: an integrated environment for supporting the design of generic automated negotiators. Comput. Intell. (2012)
20. Munim, A.: GOLEMLite: a framework for the development of agent-based applications. Master's thesis, Department of Computer Science, Royal Holloway, University of London, Sept (2013)
21. Narayanan, V., Jennings, N.R.: An adaptive bilateral negotiation model for e-commerce settings. In: Proceedings of the Seventh IEEE International Conference on E-Commerce Technology, pp. 34–41, Washington, DC, USA. IEEE Computer Society (2005)
22. Nguyen, T., Jennings, N.R.: Managing commitments in multiple concurrent negotiations. Int. J. Electron. Commer. Res. Appl. **4**, 362–376 (2005)
23. Ren, F., Zhang, M., Sim, K.M.: Adaptive conceding strategies for automated trading agents in dynamic, open markets. Decis. Support Syst. **46**(3), 704–716 (2009)
24. Stathis, K., Kakas, A.C., Lu, W., Demetriou, N., Endriss, U., Bracciali, A.: PROSOCS: a platform for programming software agents in computational logic. In: Müller, J., Petta, P. (eds.) Proceedings of the Fourth International Symposium "From Agent Theory to Agent Implementation" (AT2AI-4 – EMCSR'2004 Session M), pp. 523–528, Vienna, Austria, April "13-16" (2004)
25. Tsuruhashi, Y., Fukuta, N.: A framework for analyzing simultaneous negotiations. In: Boella, G., Elkind, E., Savarimuthu, B., Dignum, F., Purvis, M. (eds.) PRIMA 2013: Principles and Practice of Multi-Agent Systems. Lecture Notes in Computer Science, vol. 8291, pp. 526–533. Springer, Berlin (2013)
26. Williams, C.R., Robu, V., Gerding, E.H., Jennings, N.R.: Negotiating concurrently with unknown opponents in complex, real-time domains. In: 20th European Conference on Artificial Intelligence, vol. 242, pp. 834–839, Aug (2012)

Using Transfer Learning to Model Unknown Opponents in Automated Negotiations

Siqi Chen, Shuang Zhou, Gerhard Weiss and Karl Tuyls

Abstract Modeling unknown opponents is known as a key factor for the efficiency of automated negotiations. The learning processes are however challenging because of (1) the indirect way the target function can be observed, and (2) the limited amount of experience available to learn from an unknown opponent at a single session. To address these difficulties we propose to adopt two approaches from transfer learning. Both approaches transfer knowledge from previous tasks to the current negotiation of an agent to aid learn the latent behavior model of an opposing agent. The first approach achieves knowledge transfer by weighting the encounter offers of previous tasks and the ongoing task, while the second one by weighting the models learnt from the previous negotiation tasks and the model learnt from the current negotiation session. Extensive experimental results show the applicability and effectiveness of both approaches. Moreover, the robustness of the proposed approaches is evaluated using empirical game theoretic analysis.

S. Chen (✉)
School of Computer and Information Science, Southwest University,
Chongqing 400715, China
e-mail: siqichen@swu.edu.cn

S. Zhou · G. Weiss
Department of Knowledge Engineering, Maastricht University,
P.O. Box 616, 6200 MD Maastricht, The Netherlands
e-mail: shuang.zhou@maastrichtuniversity.nl

G. Weiss
e-mail: gerhard.weiss@maastrichtuniversity.nl

K. Tuyls
University of Liverpool, Ashton Street, Liverpool L69 3BX, UK
e-mail: k.tuyls@liverpool.ac.uk

© Springer International Publishing Switzerland 2016
N. Fukuta et al. (eds.), *Recent Advances in Agent-based Complex
Automated Negotiation*, Studies in Computational Intelligence 638,
DOI 10.1007/978-3-319-30307-9_11

1 Introduction

In automated negotiations, two or more autonomous agents try to come to a joint agreement in a consumer-provider or buyer-seller set-up [17]. The biggest driving force behind research into automated negotiation is arguably the broad spectrum of potential applications. Negotiation theory typically differentiates negotiation classes through their negotiation settings, for example, the number of participants on the negotiation table (e.g., bilateral or multilateral), or the number of issues being negotiated upon (e.g. whether a single or multiple issues are the subject of each placed bid). Although the contributions from our work are also applicable to multilateral negotiations, the paper concentrates on bilateral, multi-issue negotiation, simply because this makes the proposed techniques easier to explain. The interaction framework enforced in automated negotiations lends itself to the use of machine learning techniques for opponent modeling. The driving force of an (opposing) agent is governed by its hidden utility function as well as its also hidden bidding strategy. Given the opponent behavior model,[1] an agent can reach better final (or cumulative) agreement terms by exploiting this private knowledge. But learning an opposing agent's behavior model is not trivial due to the following reasons [4]:

1. the behavior model can only be observed indirectly through offers refused and counter offers made by the opposing agent;
2. the amount of encounter offers available to learn from in a single negotiation session is limited.

Transfer learning (TL) techniques are therefore adopted to alleviate the learning problems. TL is a branch of machine learning that enables the use of knowledge learned in a previous task (so called *source* task) to aid learning in a different, related new task (*target* task) [20]. In its most basic form, there exists a source and a target task where learning in the source task was already accomplished and the knowledge acquired (in whatever form) is available for use in the target task, with the underlying assumption that the source and target task can be found similar in some sense [11, 20, 25]. One of the primary goals of TL is to reach better performance in a new task (with few target data being available) by re-using gained knowledge from source task, which is ideally suited for learning settings like that of automated negotiation. More specifically, the learning of an opponent model in an ongoing negotiation (*target* task) could be benefit from transferring knowledge from previous negotiations (*source* tasks). Aiming at applying transfer learning to automated negotiation, this work contributes by:

1. proposing a generic strategy framework for agent-based negotiation;
2. modifying an instance-based transfer algorithm *TrAdaBoost* [11] for multi-issue negotiation problems, and

---

[1]Because both an agent's utility function and bidding strategy are hidden, we will often use the term behavior model to encompass both as the force governing the agents negotiating behavior.

3. modifying a model-based transfer algorithm—*ExpBoost* [22] for multi-issue negotiation problems.

The first algorithm transfers knowledge between tasks based on data instances. This approach is intuitively appealing—although the whole data from source task may not be reused directly, certain parts of the data can still be reused together with a few labeled data in the target task [20]. One of the most widely used instance transfer learning algorithm is *TrAdaBoost*, which was proposed to transfer instances between a source and a target task for classification. In the automated negotiation scenario, encountered offers from previous negotiation session are transferred to improve the learning agent's offer proposition scheme in the ongoing negotiation session. In contrast, the second algorithm is based on the model transfer learning. Model transfer approach, by means of discovering the relation between source and target models, transfers knowledge across source and target tasks. Therefore, such kind of approaches aim at retaining and applying the models learned in a single or multiple source tasks to efficiently develop an effective model for the target task, after seeing only a relatively small number of sample situations. In the automated negotiation scenario, instead of directly transfer the encountered offers from previous sessions, the learned models on historical negotiation sessions are transferred to approximate the model of the current session. It can automatically balance an ensemble of models with each trained on one known opponent.

Experiments performed on various challenging negotiation tasks show that transfer can aid target agents in improving their behaviors once encountering new opponents varying in their preference profiles and/or bidding strategies. The contributions, moreover, includes the discussion of performance improvement resulting from transfer, which opens up a few new directions of research.

The remainder of this paper is structured as follows. Section 2 underlines the problem of our research. Section 3 proposes the generic strategy framework for automated negotiation and the two transfer learning schemes. Section 4 offers extensive experimental results and a game-theoretical analysis of the proposed learning approaches. Section 5 provides related work. Section 6 identifies some important research lines outlined by our paper and concludes the work.

2 Bilateral Negotiation Problem

The automated negotiation framework adopted in this work is a basic bilateral multi-issue negotiation model as it is widely used in the agents field (e.g., [5, 7–10]). The negotiation protocol is based on a variant of the alternating offers protocol proposed in [23], where the negotiation process is limited by a certain number of rounds.

Let $I = \{a, b\}$ be a pair of negotiating agents, where i $(i \in I)$ is used to represent any of the two agents. The goal of a and b is to establish a contract for a product or service, where a contract consists of a vector of issue values. Inherent to the negotiation process is that agents a and b act in conflictive roles. To make this precise, let J be the set of issues under negotiation where j ($j \in \{1, \ldots, n\}$) is used to represent a particular issue. Contracts are tuples $O = (O_1, \ldots, O_n)$ that assign a value O_j to each issue j. A contract is said to be established if both agents agree on it. Following Rubinstein's alternating bargaining model, each agent makes, in turn, an offer in form of a contract proposal.

Each agent i decides to accept or reject a contract based on a weight vector $w^i = (w^i_1, \ldots, w^i_n)$ (also called importance vector or preference vector) that represents the relative importance of each issue $j \in \{1, \ldots, n\}$. These weights are usually normalized (i.e., $\sum_{j=1}^{n}(w^i_j) = 1$ for each agent i).

The utility of an offer for agent i is obtained by the utility function, defined as:

$$U^i(O) = \sum_{j=1}^{n}(w^i_j \cdot V^i_j(O_j)) \tag{1}$$

where w^i_j and O are as defined above and V^i_j is the evaluation function for i, mapping every possible value of issue j (i.e., O_j) to a real number. The two parties continue exchanging offers till an agreement is reached or negotiation time runs out.

3 Transfer Between Negotiation Tasks

This section details the context and the proposed algorithms for automatically transferring between two or multiple negotiation tasks within the same domain. Next, a generic framework of negotiation strategy is first given, which support both transfer and non-transfer modes. The details of the learning in each of the source and the target tasks are then presented.

3.1 The Generic Framework of Agent-Based Negotiation Strategy

In the present section, we detail the generic strategy framework. When the source task data is available, the strategy operates in the transfer learning mode to reuse knowledge from other negotiations tasks; in the other case, it works in the plain mode and decides its negotiation moves solely on the newly learnt model of the target task.

Upon receiving a counter-offer from the opponent at the time t_c, the agent records the time stamp t_c and the utility $U(O_{opp})$ of this counter-offer according to its own utility function. A small change in utility of the opponent can result in a large utility variation for the agent leading to a fatal misinterpretation of the opponent's behavior in the case of multi-issue negotiations. Therefore and in order to reduce that negative impact, the whole negotiation is divided into a fixed number (denoted as ζ) of equal intervals. The maximum utilities at each interval with the corresponding time stamps, are then provided as inputs to the learner *SPGPs* (for more details, see [24]). This also significantly scales down the computation complexity of modeling opponent so that the response speed is improved.

Then, dependent on whether the source task data is available, the agent learns the opponent differently as explained in Sects. 3.2 and 3.3, respectively. With the approximated opponent model, the agent adopts a linear concession strategy to avoid further computation load. More specifically, the optimal expected utility $\hat{u}$ (i.e., the expected maximum opponent concession) provided by the opponent model is used to set the target utility to offer at the current time t_c as follows:

$$u' = \hat{u} + (u_{\max} - \hat{u})(\hat{t} - t_c)^\alpha \tag{2}$$

where $u_{\max}$ is the maximum utility allowed in the negotiation task, $\hat{t}$ is the time when $\hat{u}$ will be reached, α the concession factor determining the way how the agent concedes, e.g., Boulware ($\alpha < 1$), Conceder ($\alpha > 1$) or Linear ($\alpha = 1$).

After the target utility u' has been chosen, the agent has to decide how to respond to the opponent's current counter-offer (this corresponds to line 14 in *Algorithm 1*). The agent first checks whether any of the following two conditions is fulfilled: (1) the utility of the latest counter-offer $U(O_{opp})$ is not smaller than u'; (2) the counter-offer has been proposed by the agent itself to its opponent at some earlier point during the ongoing negotiation process.

The agent settles the deal and the negotiation ends (line 12) if any of these two conditions is satisfied. Otherwise, the agent checks whether u' falls below the best counter-offer received so far. If this holds, then, for the consideration of efficiency, this counter-offer is proposed to the opponent. Proposing such an offer is reasonable because it tends to satisfy the expectation of the opponent. If not, the agent then constructs a new offer following a ε-greedy strategy as used in [3]. According to this strategy, a greedy offer is chosen with probability 1-ε in order to further explore the opponent behavior, and with probability ε a random offer is made (where $0 \leq \varepsilon \leq 1$). The greedy offer is chosen as follows. For a rational opponent, it is reasonable to assume that the sequence of its counter-offers is in line with its decreasing satisfaction. Thus, the more frequent and earlier a value of an issue j appears in counter-offers, the more likely it is that this issue contributes significantly to the opponent's overall utility. Formally, let $F(\cdot)$ be the frequency function defined by:

$$F^n(v_{jk}) = F^{n-1}(v_{jk}) + (1 - t)^\psi \cdot g(v_{jk}) \tag{3}$$

where the superscript of $F(\cdot)$ indicates the number of negotiation rounds, ψ is the parameter reflecting the time-discounting effect, and $g(\cdot)$ is a two-valued function whose output is 1 if the specific issue value (i.e., v_{jk}) appears in the counter-offer and 0 otherwise. The new offer to be proposed is the one whose issue values have the maximal sum of frequencies according to the frequency function and whose utility is not worse than the current target utility. In the case of a random offer, an offer whose utility is within a narrow range around u' is randomly generated and proposed.

Algorithm 1 The generic framework of negotiation strategy is described. Depending on whether source data are available, it operates in two different modes, namely, the plain and transfer mode. The idea of knowledge transfer is to re-weight instances/models from the source task such that it helps the target task agent in learning the target opponent's behaviour model Θ.

1: **Require:** different distribution (i.e., source task) labelled data sets $\mathscr{T}_D$ (if available), same distribution data set $\mathscr{T}_s$, the base learning algorithm Ξ, the maximum number of iterations N, maximum time allowed $t_{\max}$, and $t^{(k)}$ is the current time index.
2: **while** $t^{(k)} < t_{\max}$ **do**
3:    Collect time stamp and utility and add to $\mathscr{T}_S$
4:    Set n=size($\mathscr{T}_S$)
5:    **if** no source data available **then**
6:       $\Theta \Leftarrow SPGPs(\mathscr{T}_S)$
7:    **else**
8:

$$\Theta \Leftarrow \begin{cases} TrAdaBoost.Nego(\mathscr{T}_D, \mathscr{T}_S, N) & \text{instance transfer} \\ ExpBoost.Nego(\text{H}, \mathscr{T}_S, m, N) & \text{model transfer} \end{cases}$$

9:    **end if**
10:   $u' \Leftarrow setTargetU(\Theta, n)$
11:   **if** Acceptable **then**
12:      an agreement reached
13:   **else**
14:      proposes a new offer
15:   **end if**
16:   Increment time index
17: **end while**
18: Terminate the negotiation session

3.2   Learning in the Source Task

When no source tasks are at hand, the agent simply learns the current task using the plain mode of the proposed strategy. The source negotiation task starts by any side of the two parties presenting an offer describing values for the different negotiation issues. If an offer is accepted the negotiation session ends. On the other hand, if the offer is rejected the agent proposes a new offer to the other party. Then, the opponent can decide, according to her own behavior model, whether to accept or reject this new offer.

While the opponent's behavior model is unknown, it can be learned over time. The opponent model is indirectly observed from the utilities of the opponent's counter-offers: every time the opponent proposes a counter-offer, the utility of this offer is computed and added to the data set $\mathscr{T}_S = \{t^{(i)}, u^{(i)}\}_{i=1}^{t_{\max}}$, with $t^{(i)}$ representing the source task time steps running to a maximum of $t_{\max}$. The data set grows dynamically as the negotiation session continues. Every time a new instance is obtained, the model—in this case $SPGPs$—is trained anew to discover a new latent function best describing the new data set. The updated model is then used to propose a new offer to the opponent. This is achieved through the prediction probability distribution of the trained *SPGPs*. Formally, the predicted utility $u_\star$ at a new time step $t_\star$ is calculated according to the following:

$$p(u_\star|t_\star, \mathscr{D}, \bar{\mathbf{X}}) = \int p(u_\star|t_\star, \bar{\mathbf{X}}, \bar{\mathbf{f}}) p(\bar{\mathbf{f}}|\mathscr{D}, \bar{\mathbf{X}}) d\bar{\mathbf{f}} = \mathscr{N}(u_\star|\mu_\star, \sigma_\star^2), \qquad (4)$$

where

$$\mu_\star = \mathbf{k}_\star^T \mathbf{Q}_M^{-1} (\mathbf{\Lambda} + \sigma^2 \mathbf{I})^{-1} u$$
$$\sigma_\star^2 = \mathbf{K}_{\star\star} - \mathbf{k}_\star^T (\mathbf{K}_M^{-1} - \mathbf{Q}_M^{-1}) \mathbf{k}_\star + \sigma^2$$
$$\mathbf{Q}_M = \mathbf{K}_M + \mathbf{K}_{MN} (\mathbf{\Lambda} + \sigma^2 \mathbf{I})^{-1} \mathbf{K}_{NM}$$

The negotiation session ends when either an agreement is reached or the available time steps are exhausted. Finally, the opponent's utility model described by the hyper parameters of the *SPGPs* is returned for later use.

3.3   Knowledge Transfer and Target Task Learning

3.3.1   Instance-Based Transfer Approach

TrAdaBoost is originally designed for instance-based transfer learning, however it overlooks difference among source tasks. In order to make it well suited for the described negotiation setting, an extension of the standard *TrAdaBoost* algorithm is proposed to transfer instances between multiple negotiation tasks. This extended version is achieved by combining the principles of *TrAdaBoost* with the ideas of dealing with multi-task scenarios discussed in [27] and those of modifying a boosting-based classification approach for regression in [12], which together successfully results in the new regression algorithm *TrAdaBoost.Nego*. This new approach is specified in Algorithm 2.

TrAdaBoost.Nego requires two data sets as input. The first is $\mathscr{T}_D$ which represents the different distribution data set from one (or more) previous task(s) $\mathscr{T}_{D_k}$, where $\mathscr{T}_{D_k} \subseteq \mathscr{T}_D$. Since the source and the target opponent's attain their utilities from different distributions, then the different distribution data is that of the source data. Namely, $\mathscr{T}_{D_k} = \{t_k^{(i)}, u_k^{p(i)}\}_{i=1}^{|\mathscr{T}_{D_k}|}$, where $u_k^{p(i)}$ is the predicted source task's utility determined according to Eq. 4.

The second data set required by the transfer algorithm is the same distribution data set $\mathscr{T}_S$. The same distribution data set is the one from the target task having time steps as inputs and received utilities as outputs. The instances of $\mathscr{T}_S$ are attained automatically from the initial negotiation steps in the target task, where the offer proposition step depends only on the source task's knowledge. In the present case, the behavior of the opponent is monitored similar to the learning in the source task

---

Algorithm 2 TrAdaBoost.Nego ($\mathcal{T}_D$, $\mathcal{T}_S$, N)

1: **Require:** source task data sets $\mathcal{T}_D = \{\mathcal{T}_{D_1}, ..., \mathcal{T}_{D_k}\}$, the target task $\mathcal{T}_S$, the base learning algorithm $\varXi$, and the maximum number of iterations N.

2: $T = \mathcal{T}_{D_1} \bigcup ... \bigcup \mathcal{T}_{D_k} \bigcup \mathcal{T}_S$

3: $\mathbf{w} = (\mathbf{w}_{\mathcal{T}_{D_1}}, ..., \mathbf{w}_{\mathcal{T}_{D_k}}, \mathbf{w}_{\mathcal{T}_S})$

4: **Initialize:** the weight vector $\mathbf{w}^{(1)}(x_i) = \frac{1}{|T|}$, for $(x_i, c(x^{(i)})) \in T$. (Alternatively, the initial weights could be set with the user's knowledge.)

5: **for** $t = 1$ to N **do**

6:   Set $\mathbf{p}^{(t)} = \mathbf{w}^{(t)}/Z^{(t)}$ ($Z^{(t)}$ is a normalizing constant)

7:   Learn a hypothesis $h_j^{(t)} : \mathcal{X} \rightarrow \mathcal{Y}$ by calling $\varXi$ with the distribution $\mathbf{p}^{(t)}$ over the combined data set $\mathcal{T}_{D_n} \bigcup \mathcal{T}_S$.

8:   Compute the adjusted weighted prediction error of $h_i^{(t)}$ on each instance of $\mathcal{T}_S$ using:

   let $D^{(t)} = \max_{j=1}^{k} \max_{i=1}^{|\mathcal{T}_S|} |h_j^{(t)}(x^{(i)}) - c(x^{(i)})|$

   $e_{j,i}^{(t)} = \frac{|h_j^{(t)}(x^{(i)}) - c(x^{(i)})|}{D^{(t)}}$, where $(x_i, c(x^{(i)})) \in \mathcal{T}_S$

   $\varepsilon_j^{(t)} = \sum_{i=1}^{|\mathcal{T}_S|} \frac{\mathbf{w}_{\mathcal{T}_S}^{(t)}(i) e_{j,i}^{(t)}}{\sum_{q=1}^{|\mathcal{T}_S|} \mathbf{w}_{\mathcal{T}_S}^{(t)}(q)}$

9:   Choose the best hypothesis $\bar{h}^{(t)}$ such that the weighted error is minimal.

10:   Set $\beta^{(t)} = \frac{1}{2} \ln \frac{1-\varepsilon^{(t)}}{\varepsilon^{(t)}}$ and $\beta_j = \frac{1}{2} \ln(1 + \sqrt{2\ln(|\mathcal{T}_{D_j}|/N)})$, where $\mathcal{T}_{D_j} \subseteq \mathcal{T}_D$

11:   Store $\bar{h}^{(t)}$ and $\beta^{(t)}$.

12:   Update the weight vector according to:

$$
\mathbf{w}^{(t+1)} \Leftarrow \begin{cases} \mathbf{w}_{\mathcal{T}_{D_j}}^{(t)}(i) e^{-\beta_j e_{j,i}^{(t)}} & \text{for } x_i \in \mathcal{T}_{D_j} \\ \mathbf{w}_{\mathcal{T}_S}^{(t)}(i) e^{\beta^{(t)} e_{j,i}^{(t)}} & \text{for } x_i \in \mathcal{T}_S \end{cases}
$$

13: **end for**

14: **Output:** $h^{(f)}(x) = \sum_{t=1}^{N} \beta^{(t)} \bar{h}^{(t)}(x)$.

---

and the utilities of the counter-offers are computed and added to $\mathcal{T}_S$. Please note, that the number of instances from the same distribution data set, $\mathcal{T}_S$, need not be large. In fact, it suffices for to be much less than the number of instances in $\mathcal{T}_D$ for the algorithm to perform well.

Having the above data sets, the weights of each of instances are fitted according to line 12 of Algorithm 6. The principles of weight-updating mechanism remain the same as the original *TrAdaboost*. The proposed approach, however, no longer trains hypotheses by considering all source instances coming from the different distribution. Instead it generates a hypothesis for each of the source tasks, and then selects the one that appears to be the most closely related to the target. Specifically, at every iteration the i-th source task $\mathcal{T}_{D_i}$, independently from others, proposes a candidate hypothesis using a combined data set consisting of its own instances and those of the target task. The best hypothesis is then chosen such that the weighted error on $\mathcal{T}_S$ can be minimized. In this way, the impact of negative transfer caused by the imposition to transfer knowledge from a loosely related source task can be alleviated. In addition, the proposed extension adopts the way used in [12] to express each error in relation to the largest error at every iteration such that each adjusted error is still in the range [0, 1]. Although a number of loss functions are optional, our implementation employs the linear loss function as shown in line 8 because it is reported to work consistently well [21].

Once the extended version of *TrAdaBoost* algorithm fully fits the weights, the agent proposes an offer according to the predicted output of the target task function approximator in line 14 of Algorithm 2. After receiving a counter-offer, the utility of this offer is calculated and added to $\mathcal{T}_S$ so to be used in the next run of TrAdaBoost.Nego.

3.3.2   Model-Based Transfer Approach

ExpBoost is one of widely used approaches for model-based transfer learning. Different from the instance-based transfer method, *ExpBoost* makes use of prior knowledge through a group of models $\Theta_1, \Theta_2, \ldots, \Theta_B$, trained on each source data sets separately. At every iteration, it trains a new hypothesis on the weighted instances of the target task, and then combines different models in order to improve performance even if the target opponent is mostly unknown. In contrast with the instance transfer approach, only the weights of target instances are re-calculated at each run of the algorithm, and the updating rule is according to the performance of the resulting combination of hypotheses. More precisely, target instances are given more importance when they are not correctly predicted. This is because they are believed to be more informative for the next iteration and help the learning algorithm to get better estimators. In the end, the approach returns the target opponent behavior model represented by the weighted median combination of the hypotheses.

When applying the model transfer algorithm *ExpBoost* to automated negotiations introduced in this work, two issues stand out. The first one has been discussed in [21], that is, at each boosting iteration, *ExpBoost* must select between either the newly learned hypothesis on the weighted target instances or a single expert from one of the source tasks, which potentially imposes restrictions on its learning ability. The second issue relates to the one we have already discussed before—how to modify the algorithm for regression. To solve the first issue, we consider to relax the hypothesis selection constraint by allowing a linear combination of the learnt hypotheses from source tasks and the additional hypothesis trained for the target task, to minimize the error at each iteration using least squares regression. To solve the second issue, the ideas of *Adaboost.R2* [12] to deal with the way of computing the adjusted error are then incorporated into the modified version of *ExpBoost*. The modified algorithm is referred to as *ExpBoost.Nego* and present it as Algorithm 3.

Algorithm 3 ExpBoost.Nego (H, $\mathscr{T}_S$, m, N)

1: **Require:** the target task data sets $\mathscr{T}_S$ with size m, the pool of learnt model of previous tasks H = $\{\boldsymbol{\Theta}_1, \boldsymbol{\Theta}_2, \dots, \boldsymbol{\Theta}_B\}$, the base learning algorithm $\boldsymbol{\Xi}$, and the maximum number of iterations N.

2: Initialize the weight vector $\mathbf{w}^{(1)}$ with each item equally being $\frac{1}{m}$.

3: **for** $t = 1$ to N **do**

4:     Learn a hypothesis $h_{B+1}^{(t)} : \mathscr{X} \to \mathscr{Y}$ by calling $\boldsymbol{\Xi}$ with the distribution $\mathbf{w}^{(t)}$.

5:     $H_t = H \bigcup h_{B+1}^{(t)}$

6:     Find the linear combination $\bar{h}^{(t)}$ of the hypotheses from H_t, which minimizes squared error on instances of $\mathscr{T}_S$.

7:     Compute the adjusted prediction error $e^{(t)}$ of $\bar{h}^{(t)}$:
$e_i^{(t)} = \frac{|\bar{h}^t(x^{(i)}) - c(x^{(i)})|}{D^{(t)}}$, where $D^{(t)} = max_{i=1}^m |\bar{h}^t(x^{(i)}) - c(x^{(i)})|$

8:     Calculate $\varepsilon^{(t)} = \Sigma_{i=1}^m \mathbf{w}_i^{(t)} e_i^{(t)}$

9:     If $\varepsilon^{(t)} \leq 0$ or $\varepsilon^{(t)} \geq 0.5$, then stop.

10:     Let $\alpha^{(t)} = \frac{\varepsilon^{(t)}}{1 - \varepsilon^{(t)}}$

11:     Update the weight vector $\mathbf{w}_i^{(t+1)} = \frac{\mathbf{w}_i^{(t)}(\alpha^t)^{1 - e_i^{(t)}}}{Z^{(t)}}$, where $Z^{(t)}$ is a normalizing constant.

12: **end for**

13: **Output:** the weighted median of $\bar{h}^{(t)}(x)$ for $\lceil N/2 \rceil \leq t \leq N$, using the weight of $\ln(1/\alpha^{(t)})$.

---

4 Experimental Results and Analysis

The performance evaluation was done with GENIUS (General Environment for Negotiation with Intelligent multi-purpose Usage Simulation [15]). Known as a famous simulation environment, GENIUS is also adapted by the international Automated Negotiating Agents Competition (ANAC) as the official competition platform. In this simulation environment an agent can negotiate with other opposing agents representing different strategies in a specified negotiation domain, where the utility function is defined by the preference of each negotiating party. The performance of an agent can be evaluated via its utility/score achievements.

Six domains with the largest outcome space[2] are chosen from the pool of test domains created for the ANAC competitions. For negotiations in large domains, finding an offer that is acceptable to both parties becomes more of a challenge than in a small domain in the sense it is much feasible in small domains to propose a large or even the whole proportion of the possible proposals during the negotiation. These domains are therefore more complicated and computational expensive to explore, placing a big demand on the efficacy of the proposed learning schemes. The main characteristics of these domains are over viewed in Table 1 (with a descending order of the size of outcome space).

Next, we first select those source tasks that are similar to the target task for the agent to operate the proposed methods. In so doing, the difference between the behavior models in the source and target tasks is made small such that transfer can be smoothly done. Then, we run tournaments composed of a range of the state-of-

---

[2]Outcome space of a domain refers to the number of possible agreements that could be agreed upon between participants.

Table 1 Overview of test negotiation domains

Domain name	Number of issues	Number of values for each issue	Size of the outcome space
Energy	8	5	390,625
Travel	7	4–8	188,160
SuperMarket	6	4–9	98,784
Wholesaler	7	3–7	56,700
Kitchen	6	5	15,625
SmartPhone	6	4–6	12,000

the-art agents. Moreover, the order of our agents encountering other opposing agents is random, in other words, there is no any guarantee about the similarity between source and target tasks in this setting. Such tournaments are able to provide a measure of effectiveness of negotiation strategies from a realistic perspective. Finally, the empirical game theory (EGT) analysis [18] is applied to the tournaments results. Through this analysis, the strategy robustness can be well examined, which enables us to have a clear view whether the transfer learning schemes also improve the robustness of the plain strategy, or not.

4.1 Similar Source and Target Negotiation Tasks

In this set of experiments, we evaluate the performance of the proposed methods given the transferred tasks are closely related to the target task. Toward this end, we manually choose those ANAC agents as the opposing agents in source tasks, which behave in a similar way with the opponent in the target task. To illustrate the difference of the two proposed learning schemes, we implement each algorithm described in Sect. 3.3 with a distinct agent, under name of ITAgent (for instance transfer) and MTAgent (for model transfer). The plain strategy (for the case of no transfer) is implemented by PNAgent.

The results are shown in Fig. 1. Both the instance-transfer agent and model-transfer agents successfully achieved better performance than the agent using the plain strategy in the four domains. Moreover, the instance-transfer agent seemed to have a stronger negotiation power than the model-transfer agent. Another interesting observation is that the mean performance of all participants lagged far behind that of our agents. We suspect that is caused by the transfer effect since the improvement made by the transfer is at the cost of (more) benefit loss of the other party, especially when the test domains are fairly competitive.

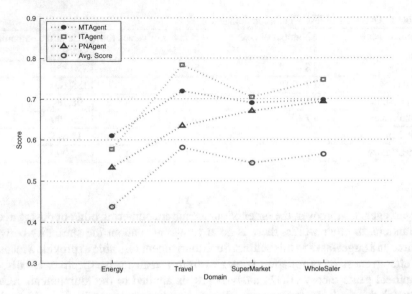

Fig. 1 Performance of the transferring and non-transferring agents in four domains when source and target tasks are similar. The average score refers to the mean score of all negotiating agents

4.2 Negotiation Tournaments

The first set of experiments, while successful, did not tell anything about how the developed strategies perform when the similarity between source and target tasks is not significant. In order to address this issue in a reasonable way, we carried out negotiation tournaments to observe agents' performance in a more realistic setting. In the tournament, each agent competes against all other agents in a random order. For each of the domains, we run a tournament consisting of the nine agents (e.g., the top three agents from respective ANAC 2012 and 2011 plus ITAgent, MTAgent and PNAgent) ten times to get results with high statistical confidence. For convenience of comparing performance across domains, normalization is adopted and done in the standard way, using the maximum and minimum raw score obtained by all agents in each domain.

According to the results given in Table 2, ITAgent and MTAgent, benefiting from the novel transfer learning schemes, took the first and second place, respectively. Both of them achieved a better score compared to the plain strategy (PNAgent), with each obtaining an improvement of 5.5 and 4.5 %. Although ITAgent performed better than the other transferring agent, the difference was small (less than 1 %). The three best agents of ANAC 2012 followed the two transferring agents, thus finishing third to fifth. The worst performers were those from ANAC 2011, whose performance was clearly below the two top agents, namely, 18 % on average.

Table 2 Overall tournament performance across all domains in descending order

Agent	Normalized score	95 % confidence interval	
		Lower bound	Upper bound
ITAgent	**0.711**	**0.694**	**0.728**
MTAgent	0.704	0.687	0.721
AgentLG	0.697	0.679	0.716
CUHKAgent	0.684	0.664	0.704
OMACagent	0.677	0.661	0.694
PNAgent	0.674	0.651	0.698
HardHeaded	0.639	0.623	0.655
IAMhaggLer2011	0.588	0.581	0.595
Gahboninho	0.572	0.558	0.584

The letter in bold of each strategy is taken as its identifier for the later EGT analysis

4.3 Empirical Game Theoretic Analysis

Till now, the performance of strategies was studied only from the traditional mean-scoring perspective. This, however, did not reveal information about the robustness of these strategies, for example, how would the results change if the players are free to switch their strategies to another one which could bring in a better individual payoff? To address robustness appropriately, the technique of empirical game theoretic (EGT) analysis [18] is applied to the tournaments results. The aim of using EGT is to search for pure Nash equilibria in which no agent has an incentive to deviate from its current strategy, or best reply cycle where there exist a set of profiles (e.g., the combination of strategies chosen by players) for which a path of deviations exists that connect them, with no deviation leading to a profile outside of the set. For brevity, the profiles in both Nash equilibria and best reply cycle are called stable states afterwards. Moreover, we consider deviations as those in [7] (the best single-agent deviation). In such a deviation one agent unilaterally changes the strategy in order to statistically improve its own profit, knowing the configuration of opponent strategies. The abbreviation for each strategy is indicated by the bold letter in Table 2. A profile (node) in the resulting EGT graph is defined by the mixture of strategies used by the players in a tournament. The first row of each node lists the three strategies, and the second row shows how many players use each strategy. The initial state of a tournament is the case where each player selects a distinct strategy. In spite of the fact that we cannot directly quantify the strategy robustness, we could rank the robustness by means of the relative sequence of a strategy being involved in stable states. Precisely, we initialize the EGT analysis with the whole set of strategies considered in our experiments. If there exist any stable state(s), the strategy attached (used) by most players is chosen as the most robust one of the current round. In the next round, we remove that winning strategy and restart the analysis to the remaining set of strategies. This process continues till all strategies are ranked or no stable state could be found in a certain round.

We visualize the results of the first three rounds under this EGT analysis in Fig. 2. As can be seen, there is a best cycle in the first round, which consists of nine stable states. AgentLG represents the most popular strategy, which is used by a total number of 52 players in all stable states of the cycle. Thus, it is the robust strategy of the first round, and also the most robust one among all candidates. Then, we proceed to the second round as indicated by the bold dotted line on the top of the figure. The only difference between the first and second round is that the robust strategy of the prior round has been excluded. The case of the second round is simpler since there exists an unique Nash equilibrium where all players attach to MTAgent. In the third round of the analysis, we again find a best cycle consisting of four states. OMACagent is voted by a number of 10 players, ITAgent attracts six players, and the other two strategies have equal less votes. As a result, OMACagent is the robust strategy selected for the third round.

The final ranking of strategy robustness is illustrated in Table 3. Surprisingly, the most stable strategy is AgentLG, even though it failed to reach the highest score achievement in the tournaments. This is because this strategy manages to demonstrate more comprehensive negotiation ability in the sense it is capable of winning more negotiations (even with a relatively smaller advantage). By contrast, ITAgent is merely in the fifth place despite being the best performer of the previous tournaments. MTAgent is more robust than the other transferring agent and the plain strategy, finishing in the second place.

The EGT analysis suggests that the proposed model-transfer learning method is robust, and seemingly has the potential of being applied in a wider range of scenarios than the instance-transfer learning method.

5 Related Work

Opponent modeling is assumed to be of key importance to performing well in automated negotiations [23]. Learning in the negotiation setting is however hampered by the limited information that can be gained about an opponent during a single negotiation session. To enable learning in this setting, various simplifying assumptions are made. For example, Lin et al. [19] introduce a reasoning model based on a decision making and beliefs updating mechanism which allows the identification of the opponent profile from a set of publicly available profiles. Another work [1] investigates online prediction of future counter-offers by using differentials, thereby assuming that the opponent strategy is defined using a combination of time- and behavior-dependent tactics [13]. Hou [16] employs non-linear regression to learn the opponent's decision function in a single-issue negotiation setting with the assumption that the opponent uses a tactic from the three tactic families introduced in [13]. Carbonneau et al. [2] use a three-layer artificial neural network to predict future counter-offers in a specific domain, but the training process requires a large

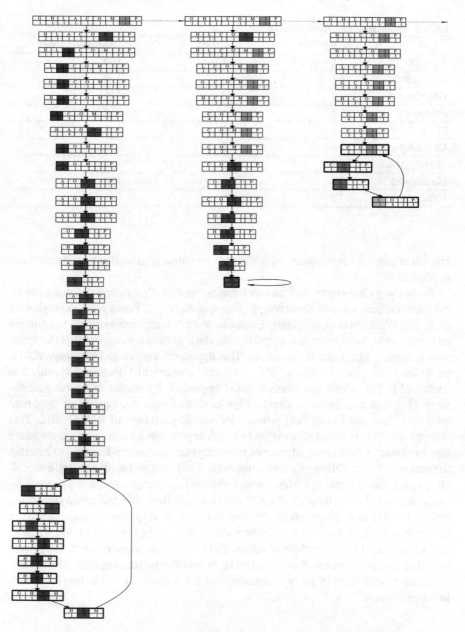

Fig. 2 The deviation analysis for the tournament results. Each node shows a strategy profile and the strategy with the highest scoring is highlighted by a background color. An *arrow* indicates statistically significant deviation from a different strategy profile. The stable states are those nodes with thinker border

Table 3 The relative robustness ranking

Agent	Ranking	Round	Attracted players
AgentLG	1	1	52
MTAgent	2	2	8
OMACagent	3	3	10
PNAgent	4	4	4
ITAgent	5	5	5
CUHKAgent	6	6	4
IAMhaggler2011	7	7	3
HardHeaded	8	8	2
Gahboninho	9	9	1

amount of previous encounters, which is time-consuming as well as computationally expensive.

Recent work has started to focus on learning opponent's strategy in a more practical situation (i.e., without those simplifying assumptions). Some good examples are [3, 6, 26]. Williams et al. [26] apply Gaussian processes regression model to optimize an agent's own concession rate by predicting the maximal concession that the opponent is expected to make in the future. This approach, known as IAMhaggler2011, made the third place in ANAC 2011. Another successful GPs-based approach is described in [3], where the authors model opponents by means of sparse pseudo-input Gaussian processes to alleviate the computational complexity of opponent modeling. Hao and Leung [14] propose the winning strategy of ANAC 2012. This strategy attempts to avoid concession as much as possible by adjusting the so-called non-exploitation time point. Moreover, it employs reinforcement-learning to increase the acceptance probability of its own proposals. Chen and Weiss [6] develop a negotiating agent, under name of OMAC, which aims at learning an opponent's strategy by analyzing its behavior through discrete wavelet transformation and cubic smoothing spline. OMAC also adapts its concession behavior in response to uncertainties in the environment. OMAC performed better than the five best agents of ANAC 2011 and was awarded the third place in ANAC 2012. Although these methods are proven useful in certain scenarios, they all suffer from insufficient training data when facing an unknown opponent in a new encounter, which results in a more or less inefficient learning process.

6 Conclusions and Future Work

This paper proposes the first robust and efficient transfer learning algorithms in negotiation tasks. The transfer technique makes use of adaptation of *TrAdaBoost* and *ExpBoost*—two well known supervised transfer algorithms—to aid learning against a new negotiating opponent. Two strategy variants were proposed. In the first the target task agent makes decisions based on weighting the instances of source and target tasks, while in the second, the agent decides its moves depending on a weighting of the source and the target models. Experimental results show the applicability of both approaches. More specifically, the results show that the proposed agents both outperform state-of-the-art negotiating agents in various negotiation domains with a considerable margin. They further demonstrate that the model transfer learning is not only a boost to the strategy performance, but also results in an improved strategy robustness.

There are several promising future directions of this work. First, the quantification of *negative transfer* is an essential next research step. Furthermore, a thorough analysis of the relationships between the agents' preferences and strategies in order to get a better understanding of transferring behavior among negotiating opponents. This understanding is also of relevance with respect to maximizing transfer stability. Last but not least, the extension of the proposed framework and strategies toward concurrent negotiations is an important issue of practical relevance that needs to be explored. In such settings, an agent is negotiating against multiple opponents simultaneously. Transfer between these tasks can serve as a potential solution for optimizing the performance in each of the negotiation sessions.

Acknowledgments We greatly appreciate the fruitful discussions with our colleagues at the Department of Knowledge Engineering, Maastricht University, especially those suggestions from Kurt Driessens, Haitham Bou Ammar and Evgueni Smirnov.

References

1. Brzostowski, J., Kowalczyk, R.: Predicting partner's behaviour in agent negotiation. In: Proceedings of AAMAS '06, pp. 355–361. ACM (2006)
2. Carbonneau, R., Kersten, G.E., Vahidov, R.: Predicting opponent's moves in electronic negotiations using neural networks. Expert Syst. Appl. **34**, 1266–1273 (2008)
3. Chen, S., Ammar, H.B., Tuyls, K., Weiss, G.: Optimizing complex automated negotiation using sparse pseudo-input Gaussian processes. In: Proceedings of AAMAS'2013, pp. 707–714. ACM (2013)
4. Chen, S., Ammar, H.B., Tuyls, K., Weiss, G.: Using conditional restricted boltzmann machine for highly competitive negotiation tasks. In: Proceedings of the 23th International Joint Conference on Artificial Intelligence, pp. 69–75, Beijing, China. AAAI Press (2013)
5. Chen, S., Hao, J., Weiss, G., Tuyls, K., Leung, H.-F.: Evaluating practical automated negotiation based on spatial evolutionary game theory. In: Lutz, C., Thielscher, M. (eds.) KI 2014: Advances in Artificial Intelligence, Lecture Notes in Computer Science, vol. 8736, pp. 147–158. Springer International Publishing (2014)
6. Chen, S., Weiss, G.: An efficient and adaptive approach to negotiation in complex environments. In: Proceedings of ECAI'2012, pp. 228–233. IOS Press (2012)
7. Chen, S., Weiss, G.: An efficient automated negotiation strategy for complex environments. Eng. Appl. Artif. Intell. **26**(10), 2613–2623 (2013)
8. Chen, S., Weiss, G.: An intelligent agent for bilateral negotiation with unknown opponents in continuous-time domains. ACM Trans. Auton. Adapt. Syst. **9**(3):16:1–16:24 (2014)

9. Chen, S., Weiss, G.: An approach to complex agent-based negotiations via effectively modeling unknown opponents. Expert Syst. Appl. **42**(5), 2287–2304 (2015)

10. Coehoorn, R.M., Jennings, N.R.: Learning on opponent's preferences to make effective multi-issue negotiation trade-offs. In: Proceedings of the 6th International Conference on Electronic Commerce, pp. 59–68, New York, NY, USA. ACM (2004)

11. Dai, W., Yang, Q., Xue, G., Yu, Y.: Boosting for transfer learning. In: Proceedings of the 24th International Conference on Machine Learning, pp. 193–200. ACM (2007)

12. Drucker, H.: Improving regressors using boosting techniques. In: Proceedings of ICML '97, pp. 107–115 (1997)

13. Faratin, P., Sierra, C., Jennings, N.R.: Negotiation decision functions for autonomous agents. Rob. Autom. Syst. **24**(4), 159–182 (1998)

14. Hao, J., Leung, H.: ABiNeS: an adaptive bilateral negotiating strategy over multiple items. In: Proceedings of the 2012 IEEE Conference on IAT, pp. 95–102 (2012)

15. Hindriks, K., Jonker, C., Kraus, S., Lin, R., Tykhonov, D.: Genius: negotiation environment for heterogeneous agents. In: Proceedings of AAMAS'2009, pp. 1397–1398 (2009)

16. Hou, C.: Predicting agents tactics in automated negotiation. In: Proceedings of the 2004 IEEE Conference on IAT, pp. 127–133 (2004)

17. Jennings, N.R., Faratin, P., Lomuscio, A.R., Parsons, S., Sierra, C., Wooldridge, M.: Automated negotiation: prospects, methods and challenges. Int. J. Group Decis. Negot. **10**(2), 199–215 (2001)

18. Jordan, P.R., Kiekintveld, C., Wellman, M.P.: Empirical game-theoretic analysis of the tac supply chain game. In: Proceedings of AAMAS'2007, pp. 1188–1195 (2007)

19. Lin, R., Kraus, S., Wilkenfeld, J., Barry, J.: Negotiating with bounded rational agents in environments with incomplete information using an automated agent. Artif. Intell. **172**, 823–851 (2008)

20. Pan, S., Yang, Q.: A survey on transfer learning. IEEE Trans. Knowl. Data Eng. **22**(10), 1345–1359 (2010)

21. Pardoe, D., Stone, P.: Boosting for regression transfer. In: Proceedings of the 27th International Conference on Machine Learning, pp. 863–870 (2010)

22. Rettinger, A., Zinkevich, M., Bowling, M.: Boosting expert ensembles for rapid concept recall. In: Proceedings Of AAAI'2006, vol. 21, pp. 464–469 (2006)

23. Rubinstein, A.: Perfect equilibrium in a bargaining model. Econometrica **50**(1), 97–109 (1982)

24. Snelson, E., Ghahramani, Z.: Sparse Gaussian processes using pseudo-inputs. In Advances In Neural Information Processing Systems, pp. 1257–1264. MIT press (2006)

25. Taylor, M.E., Stone, P.: Transfer learning for reinforcement learning domains: a survey. J. Mach. Learn. Res. **10**, 1633–1685 (2009)

26. Williams, C., Robu, V., Gerding, E., Jennings, N.: Using gaussian processes to optimise concession in complex negotiations against unknown opponents. In: Proceedings of IJCAI'2011, pp. 432–438. AAAI Press (2011)

27. Yao, Y., Doretto, G.: Boosting for transfer learning with multiple sources. In: 2010 IEEE Conference on Computer Vision and Pattern Recognition, pp. 1855–1862. IEEE (2010)

Gaussian-Based Bidding Strategies
for Service Composition Simulations

Silvia Rossi, Dario Di Nocera and Claudia Di Napoli

Abstract Service composition plays a crucial role in service–oriented computing allowing to deliver complex distributed applications obtained by aggregating autonomous and independent component services characterized by a given functionality and a Quality of Service. Automated negotiation is a viable approach to select component services according to their QoS values so to meet the end–to–end quality requirements of users requesting the application. This paper discusses the use of Gaussian probability functions to model negotiation strategies of service providers, and how the properties of these functions can be used to model multiple negotiations necessary for service composition as a single multi–issue negotiation. A numerical analysis shows comparable negotiation trends for the different representations of the service composition problem.

1 Introduction

Nowadays, Internet is evolving from a document–centric to a service–centric software infrastructure where consumers and providers engage in electronic interactions to exchange service requests and service provision. In such a scenario, it is very likely that consumers and providers play the role of buyers and sellers that initiate

Ph.D. scholarship funded by Media Motive S.r.l, POR Campania FSE 2007–2013.

S. Rossi (✉)
Dipartimento di Ingegneria Elettrica e Tecnologie dell'Informazione,
University of Naples "Federico II", Napoli, Italy
e-mail: silvia.rossi@unina.it

D. Di Nocera
Dipartimento di Matematica, University of Naples "Federico II", Napoli, Italy
e-mail: dario.dinocera@unina.it

C. Di Napoli
Istituto di Calcolo e Reti ad Alte Prestazioni—C.N.R.,
Via P. Castellino 111, 80131 Naples, Italy
e-mail: claudia.dinapoli@cnr.it

© Springer International Publishing Switzerland 2016
N. Fukuta et al. (eds.), *Recent Advances in Agent-based Complex
Automated Negotiation*, Studies in Computational Intelligence 638,
DOI 10.1007/978-3-319-30307-9_12

market–based interactions to buy and sell services. Buyers' requests are usually complex business applications that are known as Service Based Applications (SBAs), composed of an aggregation of services and requested with Quality of Services (QoSs) constraints (hard or soft) on some non–functional characteristics of the application. Examples of QoS parameters are price, response time, reliability, reputation, and so on. In order to provide QoS–aware SBAs, it is necessary to select the appropriate component services that provide the required functionalities, but at the same time also the appropriate QoS values for the considered parameters.

The problem of selecting services with QoS values, that once aggregated satisfy buyer's preferences, is an NP–hard problem, so heuristics approaches have been developed. Among these approaches, automated negotiation on QoS attributes has been adopted as a means to compose SBAs including non–functional properties. Negotiation allows to take into account the variability of the provided QoS attribute values typical of the future market of services since service providers may change these values, during the negotiation process, according to their own provision strategies. At the same time, composers may have flexible requirements on the QoS values of the required application, reflected by the strategies they adopt.

When QoS values are specified for the complete application, and not for each service, the negotiation protocols have to be more complex than traditional bilateral ones since negotiation occurs at different levels: at the level of the single services composing the SBA, and at the level of the complete application. In fact, end–to–end QoS negotiation for service composition is a multi–party negotiation in which the composer agent concurrently negotiates with multiple candidates for each component service, selecting the one that best ensures that the end–to–end QoS requirements are fulfilled. Simulating such complex interactions may be difficult and computationally expensive. Currently, the participants of QoS negotiations in SBAs adopt state–of–art strategies, drawn from research on agent–based computing. However, the use of probabilistic distributions for modeling the agent bidding behaviors may help implementing this kind of negotiations.

In this work, the negotiation strategies of service providers are based on Gaussian functions that are shown to model both one–issue and multi–issue negotiations. In addition, when using Gaussian–based strategies a single–issue negotiation for a composition of n services can be mapped to a multi–issue negotiation on n issues for one service. The use of Gaussian distributions shows that negotiation in service composition is inherently multi–issue also when a single issue is considered for the entire SBA. In addition, we show that in the case of additive issues (e.g. the price), a single–issue negotiation for a composition of services can be represented as a multi–issue negotiation for one service with the corresponding providers characterized by still mono–dimensional Gaussian distribution negotiation strategies. This results comes from the convolution properties of the Gaussian functions used to model providers' negotiation strategies, and from the independence of the service QoS values provided by different providers. So, when provider agents strategies are modeled using Gaussian–based functions, it is possible to use the same negotiation mechanism, in terms of protocol and strategies, for both single–issue, multi–issue, negotiation for service composition.

We carry out a numerical analysis to evaluate if the properties of the Gaussian functions together with the properties characterizing QoS–based service composition problems, can be used to map the problem of finding n services that, once combined, provide the overall QoS acceptable value for the complete application, in the same way as the problem of finding 1 service providing n QoS values composing an overall acceptable QoS value.

2 Background and Related Works

Negotiation usually takes place between two agents x and y willing to come to an agreement on conflicting interests. It is a bilateral interaction consisting of an alternate succession of offers and counteroffers. An offer o_x by an agent x at time t is a n–tuple $o_x(t) = (q_1, \ldots, q_n)$, where q_i is a specific value in the domain D_i of issue q_i. In order to prepare a counteroffer, an agent uses a set of tactics to generate new values for each issue [6]. Of course, both agents must be provided with strategies to formulate offers and counteroffers, and they must be equipped with algorithms to evaluate the received offers. This is done by evaluating the specific offer in terms of agent utility with respect to the offer. Hence, the utility U for an agent x is a function that depends on the specific agent x and on an offer o_i by the agent y (with $x = y$ or $x \neq y$) such as $U_x(o_y) : D_1 \times \cdots \times D_n \to [0, 1]$. The negotiation process continues either until an offer is accepted by the other agent, or one of the agents terminates the interaction (e.g., because of a deadline). An agent x accepts an offer $o_y(t)$ of y if the value of the utility the agent x obtains for that offer is greater than the utility value of the counteroffer the agent x would send back in the next iteration [9].

There are several approaches that propose negotiation mechanisms to compose services according QoS values [13, 15]. In most of these approaches negotiation occurs among a specific provider selected according to some criteria, that negotiates the values of the QoS parameters it provides with a service requester. Hence, the negotiation process is one–to–one between the service requester and the selected service provider [5, 19]. Other approaches use negotiation as a mechanism to dynamically select the appropriate service provider whose provided service best matches the requester agent QoS requirements [7]. But usually negotiation is carried out for each required service independently from the others. Attempts to propose a coordinated negotiation with all the providers of the different required services in a composition have been proposed, as in [18], but they introduce a Negotiation Coordinator that instructs the negotiation of the single component services by decomposing end–to–end QoS into local QoS requirements. On the contrary, in the present work, a coordinated negotiation mechanism is used [4], where negotiations occur concurrently with all providers of the different required services in the composition. Coordination occurs at the end of each negotiation step when the aggregated QoS values, offered by different SPs, are collectively evaluated to decide whether to accept or not a set of offers, so to take into account the dependencies among different negotiation processes.

Negotiation tractability is a fundamental requirement in order to apply negotiation mechanisms in real market applications [11]. This implies that the interaction rules have to guarantee the quick end of the process and that agents behaviors and negotiation strategies should be developed based on the assumption of bounded rather than perfect rationality [12]. One of the common requirements for a negotiation protocol is the monotonicity of the utilities of the offers $(U_x(o_x(t + 1)) \leq U_x(o_x(t)))$ as in [20]. This allows to guarantee the end of the process without a deadline: either an agreement is reached (sooner or later), or a conflict is reached in the case all agents stop to concede in utility.

3   A Multi–agent Negotiation for Service Composition

Automated negotiation has been proven to be a suitable approach to improve outcomes of market–based electronic transactions. In fact, negotiation on the QoS parameters of the services composing the application, allows to find a composition of services whose QoS values, once aggregated, meet the buyers' requirements. In service–oriented settings, negotiation protocols have to be more complex than traditional bilateral ones. For this reason, in the present work we adopt the negotiation mechanism proposed in [4], where multiple sellers for each service functionality (Abstract Services—ASs) required for the SBA concurrently negotiate with a buyer agent acting on behalf of the consumer. The negotiation protocol is based on the Iterative Contract Net Protocol (ICNP) one, that is a multi–agent iterative protocol. The negotiators are a composer agent (SC), acting on behalf of the consumer, and the different service providers (SPs) of each required SBA functionality. So, negotiation occurs at the level of the single service, and, at the same time, at the level of the whole application with the providers of the different ASs in a coordinated way. A successful negotiation consists in finding a set of interrelated non–functional values, that once aggregated meet the buyers requirements. The negotiation protocol allows the SC to negotiate separately, and hence concurrently, with all the SPs available for the same AS, and it may be iterated for a number of times (rounds) until a deadline is reached, or the negotiation is successful [4]. At each negotiation iteration, the SC issues an SBA request consisting in the specification of all component service functionality, and of the requested/preferred QoS values for the considered application attributes. The available SPs reply with offers for the service functionality they provide, specifying the QoS values of their service implementation. Negotiation takes place with all the available provider agents of each required service in the SBA, since we assume that a provider that was not promising at a given round, may become more promising in successive rounds if the values of the QoS attributes offered by the other services change. The SC can only evaluate its own utility for the aggregated QoS values of the considered parameters, and verify if they meet the requirements. It cannot issue a counter proposal since in the service–oriented scenario adopted in the present work, offers for a single functionality cannot be formulated independently from the others. In fact, if the value of one offer changes, also the value of the other offers have to change accordingly in order to meet buyer's constraints, i.e. the change of a value of a particular QoS value can impact the acceptance of the QoS values provided by the other SPs. Furthermore, there is no global shared information, since in real market of services it is unlikely that buyers have enough information about all sellers negotiation strategies to formulate single counter–offers. So, buyers can only evaluate acceptable and unacceptable offers for a complete service set. In this situation buyers and sellers are in a highly asymmetric relationship, hence symmetric protocols requiring a strong symmetry between composers and providers [8] are not appropriate.

With this assumption, the compositor agent adopts the same strategy for both single– and multi–issue negotiation consisting in simply evaluating the Euclidean distance of the aggregated value of the issues provided by the different services in the composition, from its preferred values. For a single–issue negotiation such distance is calculated in a space of dimension m, where m is the number of services required in the SBA. For a multi-issue negotiation such distance is calculated in a space of dimension $m * n$ where n is the number of issues under negotiation. The compositor accepts the offers for the complete set of services in the case the distance is equal to 0.

4 Gaussian Based Bidding Strategies

In multi–issue negotiation most approaches usually use linear utilities functions on the issues [10], while in single issue negotiation there are rarely linear but exponential or polynomial functions, classified as boulware or conceder tactics [6]. According to [6], time dependent and resource dependent strategies are two important classes of negotiation tactics in service-oriented domains, already used for modeling B2B interactions [1]. Time dependent strategies model the interactions of agents with deadlines for making deals. Usually, as the deadline approaches the agent is more willing to concede in utility. Resource dependent strategies are similar to the time dependent ones, but the domains of functions modeling the tactics are the available resources. In this case, it is necessary to evaluate the available resources according to the received requests for that resource to generate new offers.

When dealing with multi–agent multi–issue negotiation, the definition of utility functions is not a simple task. In economic literature, utility functions can be modeled as probability distributions, useful for their mathematical properties [2]. Moreover, from a computational perspective, probability distributions may have relevant properties in the negotiation process since they allow to may be used in different negotiation types with a uniform strategy approach, while simultaneously modeling both the concession strategy and the utility function. Moreover, Gaussian distributions are used in regression processes to estimate the concession of an opponent [16].

In this work, SPs use concession strategies based on Gaussian functions that model both negotiation on a single issue [4], and on multiple issues [14] where the issues represent the QoS attributes under negotiation. Such functions can be used to model issues whose offered values can decrease during the negotiation process until a *reservation value* is reached. Here, we recall the use of such distributions to model agent single– and multi––issue concession strategies, showing that their mathematical properties are useful to simulate the behavior of different agents (with different strategies) involved in the negotiation, and the composition mechanism in a simple way.

More specifically, the same utility functions and strategies used to model a negotiation on n issues for one service can be used for the case of a single issue negotiation for n services. This means that the complexity of the negotiation for SBAs depends on the number of issues, on the number of SPs, and on the number of ASs composing the SBA. In the case of a single issue negotiation for n services mono–dimensional Gaussian distributions are used, while in the case of a multi–issue negotiation for one service a multi–dimensional Gaussian distribution is used obtained by simply scaling the mono–dimensional one. Such multi–dimensions derive from the service composition process implying the composition of the QoS values of the component services to meet end–to–end constraints.

4.1 Gaussian Functions for Single Issue

A Gaussian function is characterized by its parameters μ and σ ($G(\mu, \sigma)$).

As proposed in [4], a Gaussian distribution represents the SP utility function, and at the same time the probability distribution of the offers the SP may issue. The best offer in terms of SP own utility is represented by the μ value of the Gaussian function, i.e. $U(\mu) = 1$. Note that μ also corresponds to the QoS value of the offer with the highest probability to be selected. The Gaussian standard deviation σ represents the attitude of the provider to concede during negotiation, and it determines the reservation value of the corresponding agent ($\mu - \sigma$). Hence, bigger values of σ correspond to smaller reservation values. The negotiation set of an issue is $[\mu - \sigma; \mu]$. In Fig. 1 (left), two Gaussian functions of two service providers with the same value of μ, but different σ (σ_1 and σ_2) are shown, with the corresponding offers generated during a negotiation run.

In Fig. 1 (right), the trend of the offered QoS values by the two service providers is shown on varying the negotiation rounds. At each negotiation round, an SP generates, following its probability distribution, a new utility value corresponding to a new offer. If this value is lower than the one offered in the previous round and within the negotiation set, then it proposes the new value. If this value is greater than the one offered in the previous round, or it is outside the negotiation set, the provider proposes the same value offered in the previous round. This strategy allows to simulate different and plausible behaviors of providers that prefers not having a consistent loss in utility, even though by increasing the number of negotiation rounds the probability for the provider to move towards its reservation value increases, since it will always concede.

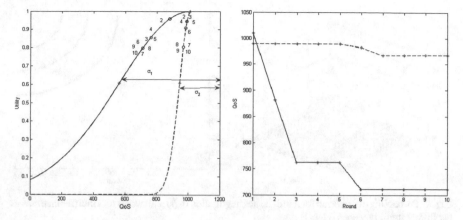

Fig. 1 Two probability distributions for a single–issue negotiation (*left*), and an example of concession rate (*right*)

4.2 Gaussian Functions for Multiple Issues

According to [14], the previous negotiation strategy for a single issue negotiation with multiple providers of different services, can be easily extended to a multi–issue negotiation. In this case, instead of the mono–dimensional Gaussian function, a multi–dimensional one is adopted to model both provider utility, and its attitude to concede.

The formula used to model a multi–variable Gaussian function is the simplest one, where we do not consider the co–variance among issues, as follows:

$$U_x(q_1, \ldots, q_n) = \prod_{i=1}^{n} \left[\exp\left(-\frac{1}{2} \left(\frac{q_i - \mu_i}{\sigma_i} \right)^2 \right) \right]$$

where, for each issue q_i, σ_i models the concession attitude when all the other $n-1$ issues are kept fixed, and μ_i is the value for the issue that corresponds to the grater value of utility ($U(q_1, \ldots, q_n) = 1$). Values of the utility function are still in the domain [0, 1], that is one dimensional (see Fig. 2 (left)).

This general representation allows to model an utility function with a non–linear dependence among different issues. Starting from this multi–dimensional Gaussian function, an utility level corresponds to an indifference curve, that includes different combinations of values, one for each issue, having the same utility value for the provider. Hence, differently from the single–issue case, here, the agent can do trade–offs between values with the same utility. Trade–offs in a continuous space may become intractable, since the problem is how to select a point in the space or to decide when to concede in utility. In fact, from the provider point of view, each point laying in the indifference curve has the same utility. Some approaches propose that the agent can make offers proposing directly all the negotiation space (with constant utility)

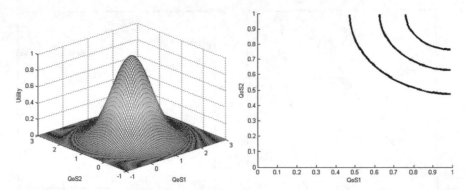

Fig. 2 Probability distribution for multi–issues negotiation (*left*) and two–issues indifference *curves* for different utility values (*right*)

[10]. However, this knowledge disclosure (with the possibility to apply optimality solution theorems), cannot be appropriate in many real SBAs scenarios. Efficient heuristics to find Pareto or quasi–Pareto optimal solutions exist, and such approaches rely on the availability of counteroffers from the other agents [10, 17], while in [3] near Pareto optimal solutions are found also for a service composition case.

In Fig. 2 (right), we show different indifference curves generated for different values of U_x. Given an utility value, the indifference curve is a projection on the issues iper–plane that models a rational and strictly convex function (properties widely applied in economics [10]). For the same value U_x a negotiation space is individuated by a section of an ellipse, that has the following equation:

$$\sum_{i=1}^{n} \frac{(q_i - \mu_i)^2}{a_i^2(U, \sigma_i)} = 1$$

where, a_i is a value that depends on the current value of the global utility, and the concession parameter σ_i. In order for a provider agent to have the rationality requirement only a single portion of such curve is allowed.

Since the distribution is built upon single concession strategies, in our approach, the concession on the utility can be evaluated by conceding on a single issue only (its marginal utility), fixing the values of all the others, and then evaluating the new corresponding utility. Hence, when conceding in utility the agent fixes $n - 1$ issues and makes a concession on a single issue selecting a value from the Gaussian distribution. If this new value is greater then the previous value (or outside the negotiation space), the agent continues to make an offer with the same utility value. However, differently from the single issue case, the actual values of the offered issues may be different (while keeping the same utility). Convexity of the utility function ensures that the agent preference on each issue is monotone when fixing the others. So, if the value increases (or decreases) the utility always decreases (or increases).

4.3 Gaussian Functions for Service Composition

Having adopted a Gaussian function to model both the strategies and the utilities of different providers, allowed to scale from a single–issue to a multi–issue negotiation by simply scaling the Gaussian dimensions.

In the case of a single issue negotiation, split among different services composing an SBA, the end–to–end requirements expressed by the user impose a relationship on the single issues, even though they are independently provided by different providers. Such relationship is expressed by an aggregation function (e.g., multiplication, sum) to obtain the global QoS of the entire composition. From a probability point of view, the composition of issues independently provided according to their probability distributions can be simulated with the probability distribution of the aggregated value. For this reason, the negotiation for service composition can be modeled as a multi–issue negotiation (with the probability distribution of the aggregated value) for a single service, and hence it is inherently multi–issue.

For example, let us consider the issue being the QoS *reliability* value. Such issue is typically aggregated through a multiplication operator. Moreover, let us assume to have two ASs in the composition and one SP for each AS. Each SP generates offers according to its Gaussian utility function (respectively, $P_1(x) = G(\mu_1, \sigma_1)$, and $P_2(y) = G(\mu_2, \sigma_2)$). The probability distribution of the aggregated offers ($c = x \cdot y$) is represented by $P(x \cdot y) = P_1(x) \cdot P_2(y)$, because the offers are independently provided, i.e., x and y are independent variables. This distribution is itself a Gaussian distribution since it is the product of two Gaussians ($P(c) = G(\mu_1, \mu_2, \sigma_1, \sigma_2)$). In Fig. 3, the probability distributions of both the composed issue for the single service (right) and of the two component issues for the two services (left) are shown.

$$(P_1(x) \cdot P_2(y)) = \exp\left[-\frac{1}{2}\left(\frac{(x - \mu_1)^2}{\sigma_1^2} + \frac{(y - \mu_2)^2}{\sigma_2^2} \right) \right]$$

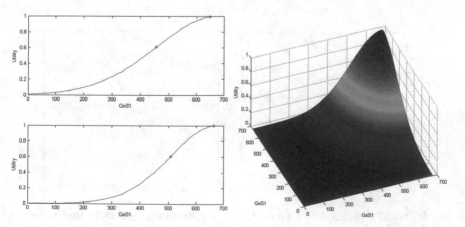

Fig. 3 Probability distributions for a single-issue negotiation with one SP for each of the two ASs (*left*), and the corresponding aggregated (multiplied) probability (*right*)

Hence, Gaussian–based utility functions and strategies used to model a negotiation on n issues for one service, which represents the composition of the SBA, are the same as the ones used to model the negotiation for the composition of n services with a single issue.

In the case there are m providers for n ASs, it is necessary to consider all the possible offer combinations in order to evaluate the end–to–end constraint satisfaction. For this reason, when mapping the probability distributions of the single services into the one of the "virtually composed" service, the number of the composed probability distributions to be considered should be exactly the same as the number of all possible offer combinations, i.e., to simulate the same problem there should be m^n "virtual providers" for the single virtually composed service.

When the issues are additive (for example the *price*), the aggregation function is a sum. The relationship among independent variables, whose values vary according to probability distributions, is modeled as a convolution of their probability distributions with additive issues. This means that the probability distribution of the sum of the two independently provided issues, is the convolution of the single probability distributions of the corresponding variables $(P(x + y) = P_1(x) \otimes P_2(y))$.

$$(P_1(x) \otimes P_2(y)) = \exp\left[-\frac{(z - (\mu_1 + \mu_2))^2}{2(\sigma_1^2 + \sigma_2^2)} \right]$$

where, z is the sum variable $(x + y)$.

Hence, since the convolution of two Gaussian functions is still a Gaussian function, the resulting function can still be used to evaluate the distance of the end–to–end QoS requirements from the aggregated QoS values received at a given negotiation round. Differently from multiplicative issues, in this case the resulting Gaussian function is still mono–dimensional. In Fig. 4, the probability distribution of both the composed issue for the single service (right) and the two component issues for the two services (left) are shown for an additive issue.

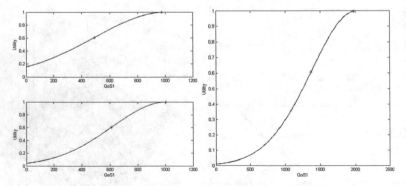

Fig. 4 Probability distributions for a single-issue negotiation with one SP for each of the two ASs (*left*), and the corresponding aggregated (summed) probability (*right*)

5 Service Composition Numerical Simulations

We carried out a set of experiments to numerically evaluate the trends of negotiation in the case of multiplicative and additive QoS issues in service composition. We also compare the trends of negotiation when the composition problem is represented in terms of the provision of a single virtually composed service. The attributes of the composed service have been composed, according to the aggregation function, that is the composition of utility functions of the components issues, and are offered following this function.

The aim of this numerical analysis is to show that the mathematical equivalence of the Gaussian distributions is also reflected in the obtained experimental negotiation trends. These trends are measured in terms of:

- the percentage of successes obtained during the negotiation (%succ),
- the number of negotiation rounds necessary to reach that success percentage (#rounds),
- the number of the corresponding exchanged messages (#mess).

These negotiation parameters are computed for an SBA request composed of 2 ASs, by varying the number of SPs (from 1 to 4) and the number of allowed negotiation rounds (from 10 to 30). The reported values are mediated on a number of 100 experiments for each considered configuration.

We first considered the case of multiplicative QoS values, such as the reliability value of the complete application that is given by the multiplication of the reliability values of its component services. For simplicity, in this case, we considered only 1 SP for each AS. The SPs probability distributions, for one negotiation run, are plotted in Fig. 5 (left), together with the offers generated at different negotiation rounds. The trends of the negotiation are compared with the same problem represented by considering a single provider whose utility function is the product of the utility functions of the two providers (one for each AS), so representing the distribution of the composed QoS values offered for the virtually composed service. An example of such distribution is reported in Fig. 5 (right), together with the composed offers

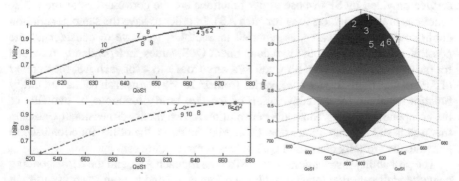

Fig. 5 Probability distributions for a single–issue (reliability) negotiation with two SPs (*left*) and an example of the aggregated probabilities (*right*)

Table 1 Negotiation trends values for multiplicative issues

Reliability						
	%succ		#rounds		#mess	
	1 SP–2 AS	1 SP_c	1 SP–2 AS	1 SP_c	1 SP–2 AS	1 SP_c
10 round	0.83 ± 0.40	0.88 ± 0.33	5.1	4.6	10	9
20 round	0.84 ± 0.37	0.97 ± 0.17	7.6	5.6	15	11
30 round	0.91 ± 0.29	0.99 ± 0.10	8.5	5.2	17	10

generated at different negotiation rounds. As discussed in Sect. 4.2, in multi–issue negotiation trade–off among issues is possible. In fact, as shown in Fig. 5 (right), in the considered negotiation run, an agreement is reached at round 7, with the last offer laying on the same indifference curve as the previous three ones.

As shown in Table 1, the percentages of successes for the two problem configurations are comparable within the error, with better values in the case of the multi–issue case, as expected. This behavior may be due to the fact that the adopted concession strategy for the composed offers (i.e., a marginal concession and a trade–off) is not the same as the one adopted for the two individual offers. In addition, in the multi–issue case, the number of offers with the same probability to be offered increases, so leading to an increased probability of success that becomes more evident by increasing the number of negotiation rounds. Accordingly, also the number of rounds necessary to reach success is lower in the multi–issue case. Finally, since in the latter case negotiation occurs only with one SP, the number of exchanged messages is lower than in the single–issue negotiation.

In the second set of experiments, we considered the case of additive QoS values, such as the price value of the complete application, that is given by the sum of the price values of its component services. Also in this case, we evaluate if the problem of finding n services that, once combined, provide an overall QoS acceptable value, is the same as the problem of finding 1 service providing n QoS values composing an overall acceptable QoS value. Hence, the trends of the single–issue negotiation are compared with the ones of the same problem represented by considering a single service provided by SPs whose utility functions are the convolutions of the utility functions of two providers (one for each AS). In order to have the same negotiation configuration in both cases, as reported in Sect. 4.3, we have to consider all the possible combinations of the two component QoS values, independently provided for the case of two services. So when SPs vary from 1 to 4 for each AS, the number of SPs in the single service case, varies from 1^2 to 4^2, where each SP has an utility function obtained by the convolution of the two utility functions of two SPs of the component services. Since the convolution of two mono–dimensional Gaussian function is still a mono–dimensional Gaussian function, the utility function used to generate the virtually composed QoS values is mono–dimensional.

The probability distributions of two SPs (one for each AS), and the corresponding convoluted distribution together with the offers generated in both cases at different

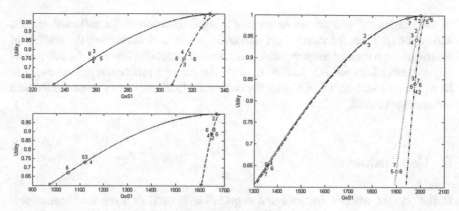

Fig. 6 Probability distributions for a single–issue (price) negotiation with two SPs (*left*) and an example of the 4 aggregated probabilities (*right*)

Table 2 Negotiation trends values for additive issues
Price

	%succ		#rounds		#mess	
	1 SP–2 AS	1 SP$_c$	1 SP–2 AS	1 SP$_c$	1 SP–2 AS	1 SP$_c$
10 round	0.13 ± 0.34	0.10 ± 0.30	9.6	9.6	19	19
20 round	0.21 ± 0.41	0.11 ± 0.31	17.6	18.7	35	37
30 round	0.23 ± 0.42	0.15 ± 0.36	25.7	27.3	51	55
	2 SP–2 AS	4 SP$_c$	2 SP–2 AS	4 SP$_c$	2 SP–2 AS	4 SP$_c$
10 round	0.34 ± 0.48	0.24 ± 0.43	8.5	8.9	34	71
20 round	0.38 ± 0.50	0.31 ± 0.46	14.7	16.2	59	129
30 round	0.43 ± 0.50	0.36 ± 0.44	21.5	25.0	86	200
	3 SP–2 AS	9 SP$_c$	3 SP–2 AS	9 SP$_c$	3 SP–2 AS	9 SP$_c$
10 round	0.43 ± 0.50	0.35 ± 0.48	8.0	8.4	48	152
20 round	0.65 ± 0.48	0.54 ± 0.50	12.0	12.4	72	223
30 round	0.70 ± 0.46	0.61 ± 0.50	15.8	18.6	95	335
	4 SP–2 AS	16 SP$_c$	4 SP–2 AS	16 SP$_c$	4 SP–2 AS	16 SP$_c$
10 round	0.61 ± 0.49	0.57 ± 0.50	7.1	6.7	57	215
20 round	0.74 ± 0.44	0.60 ± 0.49	10.6	11.7	85	374
30 round	0.78 ± 0.42	0.64 ± 0.48	14.5	15.2	112	488

rounds, for a negotiation run, are reported respectively in Fig. 6 (left) and (right). As shown in Table 2, the negotiation trends in terms of percentage of successes are comparable within the error. This is confirmed by the number of negotiation rounds for both experiments that are the same. It should be noted that the end–to–end constraints used in these experiments is set in such a way that it is difficult to reach an agreement also by increasing the number of SPs. But with this setting, a high number of negotiation failures occurs so causing a big fluctuation of the collected results, and hence an increased value of the standard deviation. Nevertheless, the number of messages exchanged during negotiation increases exponentially with the number of SPs available for each AS. In fact, in order for the two problem representations to be equivalent the number of SPs for the composed QoS increases exponentially with the number of ASs.

6 Conclusions

In the present work, software agent negotiation is used as a means to select service implementations required to deliver an SBA, by taking into account the Quality of Service that providers offer for their services, and the end–to–end QoS requirements expressed by the user requesting the application. The concession strategies adopted by service providers are modeled as Gaussian functions representing both the probability distributions of the SPs offers, and their corresponding utility values.

We showed that by modeling non linear utility functions, as well as concession strategies through the use of normal distributions, allows to deal with computational tractability requirements need by real market of services, due to their properties in scaling up in dimensions and convolutions. More specifically, with the adopted strategies, negotiation in SBAs is shown to be inherently multi–issue even in the case of a single–issue negotiation, where the multiple dimensions are given by the requirement of composing the QoS provided by the different component services to meet the end–to–end constraint. In fact, Gaussian–based utility functions used to model providers of n different services with a single QoS negotiating for reaching a successful composed QoS value, can be mapped to Gaussian–based utility functions that model providers of a single service with an already composed QoS value. The expected behavior of the negotiation trends in different but equivalent representations of a service composition problem, is supported by the reported numerical analysis.

In conclusion, the adoption of Gaussian–based strategies in negotiation for service composition allows to use the same negotiation mechanism in terms of protocols, strategies and utilities for both single issue and multi–issue negotiation, and they are easy to implement. Furthermore, the negotiation in service composition can be represented by different problem configurations that are equivalent due to the properties of the adopted Gaussian functions. These configurations can be used to carry out service composition simulations, and to evaluate which negotiation configurations to use in order to improve negotiation outcomes.

Acknowledgments The research leading to these results has received funding from the EU FP7-ICT-2012-8 under the MIDAS Project (Model and Inference Driven - Automated testing of Services architectures), Grant Agreement no. 318786, and the Italian Ministry of University and Research and EU under the PON OR.C.HE.S.T.R.A. project (ORganization of Cultural Heritage for Smart Tourism and Real-time Accessibility).

References

1. Ardagna, D., Pernici, B.: Adaptive service composition in flexible processes. IEEE Trans. Softw. Eng. **33**(6), 369–384 (2007)
2. Berhold, M.H.: The use of distribution functions to represent utility functions. Manage. Sci. **19**(7), 825–829 (1973)
3. Di Napoli, C., Di Nocera, D., Rossi, S.: Computing pareto optimal agreements in multi-issue negotiation for service composition (extended abstract). In: Proceedings of the 14th International Joint Conference on Autonomous Agents and Multiagent Systems, AAMAS'15. International Foundation for Autonomous Agents and Multiagent Systems (2015)
4. Di Napoli, C., Pisa, P., Rossi, S.: Towards a dynamic negotiation mechanism for qos-aware service markets. In: Trends in Practical Applications of Agents and Multiagent Systems, Advances in Intelligent Systems and Computing, vol. 221, pp. 9–16. Springer (2013)
5. Faratin, P., Sierra, C., Jennings, N.: Using similarity criteria to make issue trade-offs in automated negotiations. Artif. Intell. **142**(2), 205–237 (2002). International Conference on Multi-Agent Systems 2000
6. Faratin, P., Sierra, C., Jennings, N.R.: Negotiation decision functions for autonomous agents. Robot. Auton. Syst. **24**, 3–4 (1998)
7. Gimpel, H., Ludwig, H., Dan, A., Kearney, B.: Panda: Specifying policies for automated negotiations of service contracts. In: Service-Oriented Computing - ICSOC 2003. Lecture Notes in Computer Science, vol. 2910, pp. 287–302. Springer, Heidelberg (2003)
8. Hanson, J.E., Tesauro, G., Kephart, J.O., Snible, E.C.: Multi-agent implementation of asymmetric protocol for bilateral negotiations (extended abstract). In: Proceedings of the 4th ACM conference on Electronic commerce. EC '03, pp. 224–225. ACM, New York (2003)
9. Jennings, N.R., Faratin, P., Lomuscio, A.R., Parsons, S., Sierra, C., Wooldridge, M.: Automated negotiation: prospects, methods and challenges. Int. J. Group Decis. Negot. **10**(2), 199–215 (2001)
10. Lai, G., Sycara, K.: A generic framework for automated multi-attribute negotiation. Group Decis. Negot. **18**(2), 169–187 (2009)
11. Lau, R.Y.K.: Towards a web services and intelligent agents-based negotiation system for b2b ecommerce. Electronic Commerce Research and Applications **6**(3), 260–273 (2007)
12. Lomuscio, A., Wooldridge, M., Jennings, N.: A classification scheme for negotiation in electronic commerce. Group Decis. Negot. **12**(1), 31–56 (2003)
13. Paurobally, S., Tamma, V., Wooldrdige, M.: A framework for web service negotiation. ACM Trans. Auton. Adapt. Syst. **2**(4) (2007)
14. Rossi, S., Di Nocera, D., Di Napoli, C.: Normal distributions and multi-issue negotiation for service composition. In: Trends in Practical Applications of Heterogeneous Multi-Agent Systems. The PAAMS Collection, Advances in Intelligent Systems and Computing, vol. 293, pp. 1–8. Springer (2014)
15. Siala, F., Ghedira, K.: A multi-agent selection of web service providers driven by composite qos. In: Proceedings of 2011 IEEE Symposium on Computers and Communications (ISCC), pp. 55–60. IEEE (2011)
16. Williams, C.R., Robu, V., Gerding, E.H., Jennings, N.R.: Iamhaggler 2011: A gaussian process regression based negotiation agent. In: Ito, T., Zhang, M., Robu, V., Matsuo, T. (eds.) Complex Automated Negotiations: Theories, Models, and Software Competitions, Studies in Computational Intelligence, vol. 435, pp. 209–212. Springer, Heidelberg (2013)
17. Wu, M., Weerdt, M., Poutré, H.: Efficient methods for multi-agent multi-issue negotiation: allocating resources. In: Proceedings of the 12th International Conference on Principles of Practice in Multi-Agent Systems. PRIMA '09, pp. 97–112. Springer, Heidelberg (2009)

18. Yan, J., Kowalczyk, R., Lin, J., Chhetri, M.B., Goh, S.K., Zhang, J.: Autonomous service level agreement negotiation for service composition provision. Future Gener. Comput. Syst. **23**(6), 748–759 (2007)
19. Zeng, L., Benatallah, B., Ngu, A.H., Dumas, M., Kalagnanam, J., Chang, H.: Qos-aware middleware for web services composition. IEEE Trans. Softw. Eng. **30**(5), 311–327 (2004)
20. Zlotkin, G., Rosenschein, J.S.: Mechanism design for automated negotiation, and its application to task oriented domains. Artif. Intell. **86**(2), 195–244 (1996)

The Fifth Automated Negotiating Agents Competition (ANAC 2014)

Katsuhide Fujita, Reyhan Aydoğan, Tim Baarslag, Takayuki Ito and Catholijn Jonker

Abstract In May 2014, we organized the Fifth International Automated Negotiating Agents Competition (ANAC 2014) in conjunction with AAMAS 2014. ANAC is an international competition that challenges researchers to develop a successful automated negotiator for scenarios where there is incomplete information about the opponent. One of the goals of this competition is to help steer the research in the area of bilateral multi-issue negotiations, and to encourage the design of generic negotiating agents that are able to operate in a variety of scenarios. 21 teams from 13 different institutes competed in ANAC 2014. This chapter describes the participating agents and the setup of the tournament, including the different negotiation scenarios that were used in the competition. We report on the results of the qualifying and final round of the tournament.

K. Fujita (✉)
Faculty of Engineering, Tokyo University of Agriculture and Technology,
Tokyo, Japan
e-mail: katfuji@cc.tuat.ac.jp

R. Aydoğan
Computer Science Department, Özyeğin University, Istanbul, Turkey
e-mail: R.Aydogan@tudelft.nl

T. Baarslag
Agents, Interaction and Complexity Group at the University of Southampton,
Southampton, UK
e-mail: tb1m13@ecs.soton.ac.uk

T. Ito
Techno-Business Administration (MTBA), Nagoya Institute of Technolog,
Aichi, Japan
e-mail: ito.takayuki@nitech.ac.jp

C. Jonker
Man Machine Interaction Group, Delft University of Technology, Delft, Netherlands
e-mail: C.M.Jonker@tudelft.nl

© Springer International Publishing Switzerland 2016
N. Fukuta et al. (eds.), *Recent Advances in Agent-based Complex
Automated Negotiation*, Studies in Computational Intelligence 638,
DOI 10.1007/978-3-319-30307-9_13

1 Introduction

Success in developing an automated agent with negotiation capabilities has great advantages and implications. In order to help focus research on proficiently negotiating automated agents, we have organized the first automated negotiating agents competition (ANAC).

The results of the different implementations are difficult to compare, as various setups are used for experiments in ad hoc negotiation environments [7]. An additional goal of ANAC is to build a community in which work on negotiating agents can be compared by standardized negotiation benchmarks to evaluate the performance of both new and existing agents. Recently, the analysis of ANAC becomes important fields of automated negotiations in multi-agent systems [1].

In designing proficient negotiating agents, standard game-theoretic approaches cannot be directly applied. Game theory models assume complete information settings and perfect rationality [9, 10]. However, human behavior is diverse and cannot be captured by a monolithic model. Humans tend to make mistakes, and they are affected by cognitive, social and cultural factors [8]. A means of overcoming these limitations is to use heuristic approaches to design negotiating agents. When negotiating agents are designed using a heuristic method, we need an extensive evaluation, typically through simulations and empirical analysis.

We employ an environment that allows us to evaluate agents in a negotiation competition: GENIUS [7], a General Environment for Negotiation with Intelligent multi-purpose Usage Simulation. GENIUS helps facilitating the *design* and *evaluation* of automated negotiators' strategies. It allows easy development and integration of existing negotiating agents, and can be used to simulate individual negotiation sessions, as well as tournaments between negotiating agents in various negotiation scenarios. The design of general automated agents that can negotiate proficiently is a challenging task, as the designer must consider different possible environments and constraints. GENIUS can assist in this task, by allowing the specification of different negotiation domains and preference profiles by means of a graphical user interface. It can be used to train human negotiators by means of negotiations against automated agents or other people. Furthermore, it can be used to teach the design of generic automated negotiating agents.

The First Automated Negotiating Agents Competition (ANAC 2010) was held in May 2010, with the finals being run during the AAMAS 2010 conference. Seven teams had participated and three domains were used. *AgentK* generated by the Nagoya Institute of Technology team won the ANAC 2010 [2]. The Second Automated Negotiating Agents Competition (ANAC 2011) was held in May 2011, with the AAMAS 2011 conference. 18 teams had participated and eight domains were used. The new feature of ANAC 2011 was the discount factor. *HardHeaded* generated by the Delft University of Technology won the ANAC 2011 [3]. The Third Automated Negotiating Agents Competition (ANAC 2012) was held in May 2012, with the AAMAS 2012 conference. 17 teams had participated and 24 domains were used. The new feature of ANAC 2012 was the reservation value. *CUHKAgent* generated by the Chinese University of Hong Kong won the ANAC 2012 [12]. The Forth Automated Negotiating Agents Competition (ANAC 2013) was held in May 2013, with the AAMAS 2013 conference. 19 teams had participated and 24 domains were used. The new feature of ANAC 2013 was that agents can use the bidding history. *The Fawkes* generated by the Delft University of Technology won the ANAC 2013 [4].

ANAC organizers have been employing some of the new feature every year to develop the ANAC competition and the automated negotiations communities. One of the key point in achieving automated negotiation in real life is the non-linearity and size of the domains. Many real-world negotiation problems sometimes assume the nonlinear and large domains. When an automated negotiation strategy is effective to the linear function effectively, it is not always possible or desirable in the nonlinear situations [5]. In ANAC 2014, we used the constraint-based nonlinear utility function with integer issues. In addition, the domains deal with large-size domains, with outcome spaces as big as 1050 outcomes.

The remainder of this paper is organized as follows. Section 2 provides an overview over the design choices for ANAC, including the model of negotiation, tournament platform and evaluation criteria. In Sect. 3, we present the setup of ANAC 2014 followed by Sect. 4 that layouts the results of competition. Finally, Sect. 5 outlines our conclusions and our plans for future competitions.

2 Set up of ANAC

2.1 Negotiation Model

Given the goals outlined in the introduction, in this section we introduce the set-up and negotiation protocol used in ANAC. In this competition, we consider *bilateral* negotiations, i.e. negotiation between two parties. The interaction between negotiating parties is regulated by a *negotiation protocol* that defines the rules of how and when proposals can be exchanged. In the competition, we use the alternating-offers protocol for bilateral negotiation as proposed in [11], in which the negotiating parties exchange offers in turns. The alternating-offers protocol conforms with our criterion to have simplicity of rules. Moreover, it is a protocol which is widely studied and used in literature, both in game-theoretic and heuristic settings of negotiation (a non-exhaustive list includes [6, 9, 10]).

Now, the parties negotiate over a set of *issues*, and every issue has an associated range of alternatives or *values*. A negotiation *outcome* consists of a mapping of every issue to a value, and the set, Ω of all possible outcomes is called the negotiation *domain*. The domain is common knowledge to the negotiating parties and stays fixed during a single negotiation session. In addition to the domain, both parties also have privately-known preferences described by their *preference profiles* over Ω. These preferences are modeled using a utility function U that maps a possible outcomes $\omega \in \Omega$ to a real-valued number in the range [0, 1]. While the domain (i.e. the set

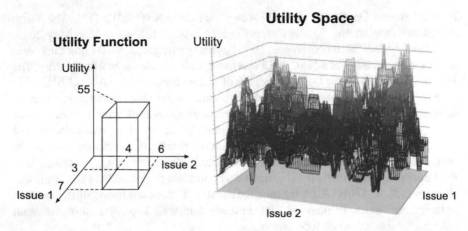

Fig. 1 Example of a nonlinear utility space

of outcomes) is common knowledge, the preference profile of each player is private information. This means that each player has only access to its own utility function, and does not know the preferences of its opponent.[1] Moreover, we use the term *scenario* to refer to the domain and the pair of preference profiles (for each agent) combined.

In ANAC 2014, we focus on *nonlinear* domains settings with a finite set of integer values per issue. An agent's utility function, in the formulation, is described in terms of constraints. There are l constraints, $c_k \in C$. Each constraint represents a region in the contract space with one or more dimensions and an associated utility value. In addition, c_k has value $v_a(c_k, \mathbf{s})$ if and only if it is satisfied by contract $\mathbf{s}$. Every agent has its own, typically unique, set of constraints. An agent's utility for contract $\mathbf{s}$ is defined as the weighted sum of the utility for all the constraints it satisfies, i.e., as $u_a(\mathbf{s}) = \sum_{c_k \in C, \mathbf{s} \in x(c_k)} v_a(c_k, \mathbf{s})$, where $x(c_k)$ is a set of possible contracts (solutions) of c_k. This expression produces a "bumpy" nonlinear utility function with high points where many constraints are satisfied and lower regions where few or no constraints are satisfied. This represents a crucial departure from previous efforts on multi-issue negotiation, where contract utility is calculated as the weighted sum of the utilities for individual issues, producing utility functions shaped like flat hyperplanes with a single optimum.

Figure 1 shows an example of a utility space generated via a collection of binary constraints involving Issues 1 and 2. In addition, the number of terms is two. The example, which has a value of 55, holds if the value for Issue 1 is in the range [3, 7] and the value for Issue 2 is in the range [4, 6]. The utility function is highly nonlinear with many hills and valleys. This constraint-based utility function representation

---

[1] We note that, in the competition each agent plays *both* preference profiles, and therefore it would be possible in theory to learn the opponent's preferences. However, the rules explicitly disallow learning *between* negotiation sessions, and only *within* a negotiation session. This is done so that agents need to be designed to deal with unknown opponents.

allows us to capture the issue interdependencies common in real-world negotiations. The constraint in Fig. 1, for example, captures the fact that a value of 4 is desirable for issue 1 if issue 2 has the value 4, 5 or 6. Note, however, that this representation is also capable of capturing linear utility functions as a special case (they can be captured as a series of unary constraints). A negotiation protocol for complex contracts can, therefore, handle linear contract negotiations.

Finally, we supplement it with a deadline, reservation value and discount factors. The reasons for doing so are both pragmatic and to make the competition more interesting from a theoretical perspective. In addition, as opposed to having a fixed number of rounds, both the discount factor are measured in *real time*. In particular, it introduces yet another factor of uncertainty since it is now unclear how many negotiation rounds there will be, and how much time an opponent requires to compute a counter offer. In ANAC 2014, the discount factors and reservation value depend on the scenario, but the deadline is set to 3 min. The implementation of discount factors in ANAC 2014 is as follows.

A negotiation lasts a predefined time in seconds *(deadline)*. The time line is normalized, i.e.: time $t \in [0, 1]$, where $t = 0$ represents the start of the negotiation and $t = 1$ represents the deadline. When agents can make agreements in the deadline, the individual utilities of each agent are the *reservation value*. Apart from a deadline, a scenario may also feature *discount factors*. Discount factors decrease the utility of the bids under negotiation as time passes. Let d in $[0, 1]$ be the discount factor. Let t in $[0, 1]$ be the current normalized time, as defined by the timeline. We compute the discounted utility U_D^t of an outcome ω from the undiscounted utility function U as follows:

$$U_D^t(\omega) = U(\omega) \cdot d^t \tag{1}$$

At $t = 1$, the original utility is multiplied by the discount factor. Furthermore, if $d = 1$, the utility is not affected by time, and such a scenario is considered to be undiscounted.

2.2 Running the Tournament

As a tournament platform to run and analyse the negotiations, we use the GENIUS environment (General Environment for Negotiation with Intelligent multi-purpose Usage Simulation) [7]. GENIUS is a research tool for automated multi-issue negotiation, that facilitates the design and evaluation of automated negotiators' strategies. It also provides an easily accessible framework to develop negotiating agents via a public API. This setup makes it straightforward to implement an agent and to focus on the development of strategies that work in a general environment.

GENIUS incorporates several mechanisms that aim to support the design of a general automated negotiator. The first mechanism is an analytical toolbox, which provides a variety of tools to analyse the performance of agents, the outcome of the negotiation and its dynamics. The second mechanism is a repository of domains and utility functions. Lastly, it also comprises repositories of automated negotiators. In addition, GENIUS enables the evaluation of different strategies used by automated agents that were designed using the tool. This is an important contribution as it allows researchers to empirically and *objectively* compare their agents with others in different domains and settings.

The timeline of ANAC 2014 consists of two phases: the qualifying round and the final round. The domains and preference profiles used during the competition are not known in advance and were designed by the organizers. An agent's success is measured using the evaluation metric in all negotiations of the tournament for which it is scheduled.

First, a *qualifying round* was played in order to select the finalists from the 19 agents that were submitted by the participating teams (2 agents were disqualified from the trial tests). Since there were 19 agents, which each negotiate against 18 other agents, in the different domains, a total pair-wise tournament in the qualifying round is impossible. Therefore, 19 agents was divided to three groups (pools) randomly, and the best three agents in social welfare and individual utility in each pool proceed to the final round. It took two weeks to finish the all pools of the qualifying round. In ANAC 2014, we didn't allow the updating agents between the qualifying round and the final round.

The final round was played among the agents that achieved the best scores (individual utility and social welfare) in each pool during qualifying. The domains and preference profiles are same as the qualifying round. The entire pairwise matches played among 10 agents, and the final ranking of ANAC 2014 was decided. In the final, a single tournament consists of $10 \times 9/2 \times 2 \times 12$ (*domains*) = 1080 negotiation sessions.[2] Again, each single tournament was repeated five to prohibit the learning from the previous tournaments. To reduce the effect of variation in the results, the tournament was repeated 5 times, and the final score means the average of the five trials.

3 Competition Domains and Agents

3.1 Scenario Descriptions

The ANAC is aimed towards modeling multi-issue negotiations in uncertain, open environments, in which agents do not know what the preference profile of the opponent is. The various characteristics of a negotiation scenario such as size, number of issues, opposition, discount factor and reservation value can have a great influence on

---

[2]The combinations of 10 agents are $10 \times 9/2$, however, agents play each domain against each other twice by switching the roles.

Table 1 The domains used in ANAC 2014

ID	Number of issues	Size	Discount factor	Reservation value
1	10	10^{10}	None	None
2	10	10^{10}	0.50	None
3	10	10^{10}	None	0.75
4	10	10^{10}	0.50	0.75
5	30	10^{30}	None	None
6	30	10^{30}	0.50	None
7	30	10^{30}	None	0.75
8	30	10^{30}	0.50	0.75
9	50	10^{50}	None	None
10	50	10^{50}	0.50	None
11	50	10^{50}	None	0.75
12	50	10^{50}	0.50	0.75

the negotiation outcome. Therefore, we generated three types of domains and profiles in the competition because the nonlinear domains are generated easily. Especially, in the qualifying round and final round, we used all 12 scenarios by allocating different discount factors and reservation values to three types of domains and profiles. In other words, they have vary in terms of the number of issues, the number of possible proposals, the opposition of the preference profiles and the mean distance of all of the points in the outcome space to the Pareto frontier (see Table 1). The shapes of the outcome spaces of each scenario are represented graphically in Fig. 2.

In generating the domains for the competition, the agents can negotiate across a variety of negotiation scenarios. In addition, the challenge of ANAC 2014 is on negotiating with nonlinear utility functions as well as dealing with large-scale outcome space. Up to this year, additive utility functions have been employed to represent agents' preferences in ANAC. Although this type of functions is compact and easy to process, they cannot represent preferential interdependencies where in many real life problem we have. Therefore, we employed the constrains-based and large-sized utility functions in ANAC 2014.

3.2 Agent Descriptions

ANAC 2014 had 21 agents, registered from 13 institutes from 8 countries: GWDG, Germany; Sun Yat-sen University, China; Bar-Ilan University, Israel; University of Isfahan, Iran; Tokyo University of Agriculture and Technology, Japan; IIIA-CSIC Barcelona, Spain; Shizuoka University, Japan; Royal Holloway University of London, U.K.; University of Electro-Communications, Japan; Delft University of Technology, The Netherlands; Hunan University, China and Nagoya Institute of Technology, Japan ($\times 9$).

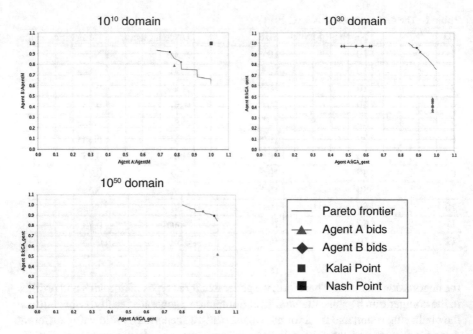

Fig. 2 Acceptance outcome space

The final round in ANAC 2014 had ten teams from eight different universities, as listed in Table 2. They are the winners of the qualifying round. In the rest of the chapter in this book, we provide sections of the individual strategies of the ANAC 2014 finalists based on descriptions of the strategies provided by the teams.

4 Competition Results

4.1 Qualifying Round

First, a *qualifying round* was played in order to select the finalists from the 19 agents that were submitted by the participating teams (2 agents were disqualified from the trial tests) 19 agents was divided to three groups (pools) randomly, and the best three agents in social welfare and individual utility in each pool proceeded to the final round. Each tournament wasn't repeated to prohibit the learning from the previous tournaments.

In order to complete such an extensive set of tournaments within a limited time frame, we used five high-spec computers, made available by Nagoya Institute of Technology and Tokyo University of Agriculture and Technology. Specifically, each of these machines contained an *Intel Core i7* CPU, at least 16GB of DDR3 memory, and a hard drive with at least 2TB of capacity.

Table 2 Team members and agent names in ANAC 2014

No.	Team members	Affliction	Agent name
1	Eden Shalom Erez	Bar Ilan University	DoNA
	Inon Zuckerman	Ariel University	
2	Farhad Zafari	University of Isfahan	BraveCat
	Faria Nasiri Mofakham		
3	Shinji Kakimoto	Tokyo University of Agriculture and Technology	kGAgent
	Katsuhide Fujita		
4	Motoki Sato	Nagoya Institute of Technology	WhaleAgent
5	Makoto Niimi	Nagoya Institute of Technology	AgentM
6	Dave de Jonge	IIIA-CSIC Barcelona	Gangster
7	Yuichi Enoki	Nagoya Institute of Technology	E2Agent
8	Yoshiaki Kadono	Shizuoka University	AgentYK
9	Satoshi Takahashi	University of Electro-Communications	Sobut
10	Balint Szollosi-Nagy	Delft University of Technology	Group2Agent
	Marta Skarzynska		
	David Festen		
11	Edwin Yaqub	Gesellschaft fur wissenschaftliche Datenverarbeitung (GWDG)	AgentQuest
12	Naiqi Li	Sun Yat-sen University	Flinch
	Zhansong Li		
13	Akiyuki Mori	Nagoya Institute of Technology	Atlas
14	Yoshitaka Torii	Nagoya Institute of Technology	agentTRP
15	Shota Morii	Nagoya Institute of Technology	Aster
16	Yoshihito Sano	Shizuoka University	AgentTD
	Tomohiro Ono		
	Takumi Wakasa		
17	Bedour Alrayes	Royal Holloway University of London	Simpatico
	Paulo Ricca		
	Ozgur Kafali		
	Kostas Stathis		
18	Taniguchi Keiichiro	Nagoya Institute of Technology	ArisawaYaki
19	Kimata	None	Simple ANAC 2013

Agent Name	Mean (Individual)	Rank (Individual)	Mean (Social welfare)	Rank (Social)
E2Agent	0.60449771	1	1.467013776	1
GROUP2Agent	0.569022057	2	1.309827507	2
kGA_gent	0.567855409	3	1.253293459	4
Sobut	0.514388859	4	1.25826658	3
ArisawaYaki	0.502270746	5	1.216825393	6
Simple ANAC2013	0.498502294	6	1.219932496	5

Fig. 3 Average scores of each agent in the qualifying round (pool1)

Agent Name	Mean (Individual)	Rank (Individual)	Mean (Social welfare)	Rank (Social)
Gangster	0.694014347	1	1.596774909	2
WhaleAgent	0.682003332	2	1.606191324	1
AgentYK	0.659956126	3	1.59018266	3
Flinch	0.649326182	4	1.573767682	4
AgentQuest	0.639303883	5	1.472990004	6
Simpaico	0.595949075	6	1.509024094	5

Fig. 4 Average scores of each agent in the qualifying round (pool2)

Agent Name	Mean (Individual)	Rank (Individual)	Mean (Social welfare)	Rank (Social)
DoNA	0.668464329	1	1.285703724	2
AgentM	0.542950221	2	1.28268408	3
BraveCat v0.3	0.518940747	3	1.422961239	1
AgentTRP	0.484535552	4	1.119857699	5
Aster	0.479403688	5	1.112286696	6
AgentTD	0.464952079	6	1.168321409	4
Atlas	0.410946126	7	0.947732281	7

Fig. 5 Average scores of each agent in the qualifying round (pool3)

Figures 3, 4 and 5 show the results of each agent in the qualifying round (pool1, pool2, and pool3). The finalists are selected from all pools by considering the individual utilities and social welfare. The individual utility means the average of utility of the individual agent in the tournaments. The social welfare means the average of the sum of utilities of two agents in the tournaments. As figures showing, the best three or four agents are selected by considering the individual utility and social welfare. As a results, *kGAgent, E2Agent, GROUP2Agent, Sobut* are selected as finalists from the pool1; *Gangster, WhaleAgent, AgentYK* are selected as finalists from pool2; *DoNA, AgentM, BraveCat* are selected as finalists from pool3. They are the best three in each pool considering the individual utility or the social welfare.

4.2 Final Round

The final round consisted of 10 agents that were selected from the qualifying round. For each pair of agents, under each preference profile, we ran a total of some negotiations. By averaging over all the scores (individual utility and social welfare) achieved by each agent (against all opponents and using all preference profiles), the final ranking were decided based on their average scores. Formally, the average score $U_\Omega(p)$ of agent p in scenario Ω is given by:

$$U_\Omega(p) = \frac{\sum_{p' \in P, p \neq p'} U_\Omega(p, p') + U_\Omega(p', p)}{2 \cdot (|P| - 1)} \qquad (2)$$

where P is the set of players and $U_\Omega(p, p')$ is the utility achieved by player p against player p' when player p is under the side A of Ω and player p' is under the side B of Ω. For the final round, we matched each pair of finalist agents, under each preference profile, a total of 5 times.

It is notable that *AgentM* was the clear winner of the both categories (see Tables 3 and 4). However, the differences in utilities between many of the ranked strategies are small, so several of the agents were decided the ranking by a small margin. Finally, the first places in the individual utility and social welfare categories were awarded

Table 3 Tournament results in the final round (Individual utility)

Rank	Agent	Score	Variance
1	Agent M	0.754618239	3.12×10^{-5}
2	DoNA	0.742245035	9.31×10^{-6}
3	Gangster	0.740674889	3.49×10^{-6}
4	WhaleAgent	0.710740252	3.90×10^{-5}
5	GROUP2Agent	0.708401404	6.38×10^{-5}
6	E2Agent	0.703955008	2.85×10^{-5}
7	kGAgent	0.676595111	5.02×10^{-5}
8	AgentYK	0.666450943	2.38×10^{-5}
9	BraveCat	0.661940343	2.84×10^{-5}
10	ANAC2014Agent	0.627684701	1.71×10^{-5}

Table 4 Tournament results in the final round (Social welfare)

Rank	Agent	Score	Variance
1	Agent M	1.645412137	4.12×10^{-5}
2	Gangster	1.627451908	1.21×10^{-5}
3	E2Agent	1.608936143	1.39×10^{-5}
4	WhaleAgent	1.603199277	3.55×10^{-5}
5	AgentYK	1.569877186	1.16×10^{-4}
6	GROUP2Agent	1.56154598	8.46×10^{-5}
7	BraveCat v0.3	1.545384774	3.11×10^{-5}
8	DoNA	1.473686528	3.89×10^{-5}
9	ANAC2014Agent	1.469972333	1.12×10^{-4}
10	kGAgent	1.463168543	4.32×10^{-4}

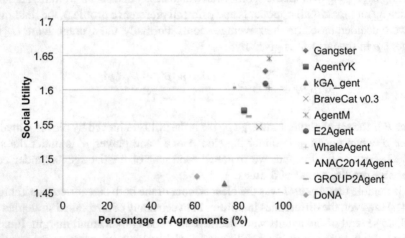

Fig. 6 Plotting graph between the percentage of agreements and the social welfare (A correlation coefficient = 0.8735)

to AgentM agent ($600); The second place in the individual category was awarded to the DoNA ($200); The second place in the social welfare was awarded to the Gangster ($200).

In more detail, we can analyze the relationships between the social welfare and other measures. As figures and showing, the percentage of agreements and the pareto distance are important features of obtaining the high social welfare. Especially, the correlation coefficient of the percentage of agreements is about 1.0 and the average of pareto distance is about −1.0. In other words, the effective strategy of obtaining the social welfare is that finding the pareto frontiers with the high percentage of agreements (Figs. 6 and 7).

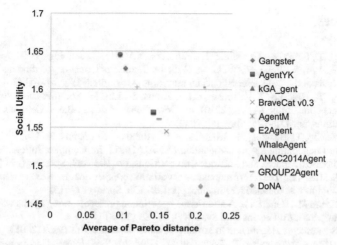

Fig. 7 Plotting graph between the average of pareto distance and the social welfare (A correlation coefficient $= -0.9994$)

5 Conclusion

This paper describes the fifth automated negotiating agents competition. Based on the process, the submissions and the closing session of the competition we believe that our aim has been accomplished. Recall that we set out for this competition in order to steer the research in the area bilateral multi-issue closed negotiation. 21 teams have participated in the competition and we hope that many more will participate in the following competitions.

ANAC also has an impact on the development of GENIUS. We have released a new, public build of GENIUS[3] containing all relevant aspects of ANAC. In particular, this includes all domains, preference profiles and agents that were used in the competition. This will make the complete setup of ANAC available to the negotiation research community. Not only have we learnt from the strategy concepts introduced in ANAC, we have also gained understanding in the correct setup of a negotiation competition. The joint discussion with the teams gives great insights into the organizing side of the competition.

To summarize, the agents developed for ANAC will proceed the next step towards creating autonomous bargaining agents for real negotiation problems. We plan to organize the next ANAC in conjunction with the next AAMAS conference.

Acknowledgments The authors would like to thank the team of masters students at Nagoya Institute of Technology, Japan for their valuable help in the organization of the ANAC 2014 competition.

---

[3] http://ii.tudelft.nl/genius.

References

1. Baarslag, T., Fujita, K., Gerding, E.H., Hindriks, K.V., Ito, T., Jennings, N.R., Jonker, C.M., Kraus, S., Lin, R., Robu, V., Williams, C.R.: Evaluating practical negotiating agents: results and analysis of the 2011 international competition. Artif. Intell. **198**, 73–103 (2013)
2. Baarslag, T., Hindriks, K.V., Jonker, C.M., Kraus, S., Lin, R.: The first automated negotiating agents competition (ANAC 2010). In: New Trends in Agent-Based Complex Automated Negotiations, pp. 113–135. Springer (2012)
3. Fujita, K., Ito, T., Baarslag, T., Hindriks, K.V., Jonker, C.M., Kraus, S., Lin, R.: The second automated negotiating agents competition (ANAC 2011). In: Complex Automated Negotiations: Theories, Models, and Software Competitions, pp. 183–197. Springer (2013)
4. Gal, Y.K., Ilany, L.: The fourth automated negotiation competition. In: Next Frontier in Agent-Based Complex Automated Negotiation, pp. 129–136. Springer (2015)
5. Ito, T., Klein, M., Hattori, H.: A multi-issue negotiation protocol among agents with nonlinear utility functions. Multiagent Grid Syst. **4**(1), 67–83 (2008)
6. Kraus, S.: Strategic Negotiation in Multiagent Environments. MIT Press (2001)
7. Lin, R., Kraus, S., Baarslag, T., Tykhonov, D., Hindriks, K.V., Jonker, C.M.: Genius: an integrated environment for supporting the design of generic automated negotiators. Comput. Intell. **30**(1), 48–70 (2014)
8. McKelvey, R.D., Palfrey, T.R.: An experimental study of the centipede game. Econometrica **60**(4), 803–36 (1992)
9. Osborne, M., Rubinstein, A.: Bargaining and Markets. Economic Theory, Econometrics, and Mathematical Economics. Academic Press (1990)
10. Osborne, M.J., Rubinstein, A.: A Course in Game Theory. MIT Press Books. The MIT Press (1994)
11. Rubinstein, A.: Perfect equilibrium in a bargaining model. Econometrica **50**(1), 97–109 (1982)
12. Williams, C.R., Robu, V., Gerding, E.H., Jennings, N.R.: An overview of the results and insights from the third automated negotiating agents competition (ANAC 2012). In: Novel Insights in Agent-based Complex Automated Negotiation, pp. 151–162. Springer (2014)

GANGSTER: An Automated Negotiator Applying Genetic Algorithms

Dave de Jonge and Carles Sierra

Abstract Negotiation is an essential skill for agents in a multiagent system. Much work has been published on this subject, but traditional approaches assume negotiators are able to evaluate all possible deals and pick the one that is best according to some negotiation strategy. Such an approach fails when the set of possible deals is too large to analyze exhaustively. For this reason the Annual Negotiating Agents Competition of 2014 has focused on negotiations over very large agreement spaces. In this paper we present a negotiating agent that explores the search space by means of a Genetic Algorithm. It has participated in the competition successfully and finished in 2nd and 3rd place in the two categories of the competition respectively.

1 Introduction

In this paper we present the agent that we have developed for the Annual Negotiating Agents Competition 2014 (ANAC'14). Our agent is called GANGSTER, which stands for Genetic Algorithm NeGotiator Subject To alternating offERs.

Theoretical work on negotiations has been published as early as 1950 [9]. The study of negotiations from an algorithmic point of view however is a much more recent topic. In [5, 6] the authors propose a strategy that amounts to determining for each time t which utility value should be demanded from the opponent (the *aspiration level*). However, they do not take into account that one first needs to find a deal that indeed yields that aspired utility level. They simply assume that such a deal always exists, and that the negotiator can find it without any effort.

To overcome that shortcoming, work on negotiations on large spaces was done in [7, 8]. They chose a model in which utility functions are nonlinear over an abstract vector space that represents the set of possible deals, but such that for any given deal

D. de Jonge (✉) · C. Sierra
IIIA-CSIC, Campus de la UAB S/n, Bellaterra, Catalonia, Spain
e-mail: davedejonge@iiia.csic.es

C. Sierra
e-mail: sierra@eiiia.csic.es

© Springer International Publishing Switzerland 2016
N. Fukuta et al. (eds.), *Recent Advances in Agent-based Complex Automated Negotiation*, Studies in Computational Intelligence 638,
DOI 10.1007/978-3-319-30307-9_14

x its utility value can still be calculated by solving a linear equation. A similar model is also used for the ANAC'14 competition, as we will see in Sect. 2.2.

The idea of large spaces with non-linear functions was carried even further by ourselves in [2, 3], where we introduced the Negotiating Salesmen Problem: a negotiation scenario in which not only the number of possible deals is very large, but, given a deal it is also very hard to determine its utility value as it requires solving a Traveling Salesman Problem. In order to tackle that domain we applied a Branch & Bound algorithm adapted to negotiations.

In this paper however, we apply Genetic Algorithms to perform the search. Furthermore, we present a new acceptance strategy that introduces the concept of reproposing, and we present a new bargaining strategy that not only depends on an aspiration level, but also on the distance between a new proposal and the proposals that have been made before.

2 Setup of the Competition

Before explaining the algorithm, we first explain how the ANAC'14 competition was set up. We use the notation $[a, b]$, where a and b are integers, to denote the set of integers z such that $a \leq z \leq b$. We use α_1 to denote our agent and α_2 to denote its opponent. Furthermore, we use $\mathcal{H}_{1 \rightarrow 2}(t)$ to denote the set of proposals made by α_1 until time t, and $\mathcal{H}_{2 \rightarrow 1}(t)$ to denote the set of proposals made by α_2 until time t.

2.1 The Protocol

In the competition each agent engaged in several negotiation sessions. In each session two agents were paired to negotiate against each other. Each session would finish as soon as the agents made an agreement, or when the deadline of 180 s had passed. The agents had to negotiate according to the alternating offers protocol [10]. One of the two agents begins. That agent may pick one deal x from the agreement space Agr_n (see Sect. 2.2) and propose it to the other. The second agent may then either accept that proposal, in which case the session finishes, or may pick another deal from the agreement space and propose it to the first agent. This then continues: agents alternately take turns, and in each turn the agent whose turn it is may make a new proposal or accept the last proposal made by the other agent. When it is agent α_1's turn agent α_2 cannot do anything, and vice versa. Furthermore, an agent may take as much time as he likes before making the next proposal (or accepting the previous proposal). Therefore, the agent's decision is not only *what* deal to propose (or accept) but also *when* to propose. More precisely: when an agent finds a potential deal to propose it should determine whether it will propose that deal or whether it should continue searching for a better deal.

If the deadline passes without any agreement having been made each agent receives a certain number of points, called its *reservation value*. Otherwise, each agent receives the amount of points equal to its utility value $f_i(x)$ (see Sect. 2.2) for the deal x they agreed upon.

2.2 The Agreement Space

In each session the space of possible deals that can be made by the agents (the *agreement space*) Agr_n is represented by an n-dimensional vector space, where n varies per session and can be as high as 50. For each vector entry there are 10 possible values.

$$Agr_n = [0, 9]^n$$

For a vector $x \in Agr_n$ we use the notation x_j to denote its j-th entry.

$$x = (x_1, x_2, \ldots x_n)$$

Definition 1 A **rectangular subspace** s is a subset of Agr_n such that for each $j \in [1, n]$ there are two integers $a_j, b_j \in [0, 9]$ with:

$$x \in s \quad \text{iff} \quad \forall j \in [1, n] : x_j \in [a_j, b_j]$$

Definition 2 A **constraint** is c a pair (s_c, v_c) where s_c is a rectangular subset of Agr_n and v_c is a real number. We say a deal $x \in Agr_n$ **satisfies** a constraint c, iff $x \in s_c$. The **characteristic function** f^c of a constraint c is defined as:

$$f^c(x) = \begin{cases} v_c & \text{if } x \in s_c \\ 0 & \text{if } x \notin s_c \end{cases}$$

For each agent $\alpha_i \in \{\alpha_1, \alpha_2\}$ there is a set of constraints C_i, that determine the agent's utility function f_i according to:

$$f_i(x) = \sum_{c \in C_i} f^c(x)$$

Both sets of constraints however, remain hidden for both agents, so an agent cannot directly calculate its own utility values. Instead, each agent α_i has access to an 'oracle' that, for any given deal $x \in Agr_n$ returns its corresponding utility value $f_i(x)$. The agents cannot know anything about their opponents' utility functions (i.e. agent α_1 can request $f_1(x)$ from its oracle, but not $f_2(x)$). Note that, in principle, an agent can request the value of each deal in the agreement space. However, since the agreement spaces are extremely large (consisting of up to 10^{50} deals) this is obviously infeasible, so the agents can only explore a tiny fraction of the agreement space. In the rest of this paper, whenever we use the term 'utility' without specifying an agent, we mean the utility function f_1 of our agent α_1.

3 Overview of the Algorithm

Let us first give a global overview of the algorithm before we go into more detail on each of its steps. Each turn the algorithm takes the following steps:

1. Calculate the *aspiration value* and *max distance* (Algorithm 1, lines 1–2).
2. Decide whether to accept the previous offer made by the opponent (Algorithm 1, lines 3–11).
3. Apply a *Global* genetic algorithm to sample the agreement space, and store the 10 proposals with highest utility (Algorithm 1, lines 12–13).
4. Apply a *Local* genetic algorithm to sample the agreement space, and store the 10 proposals with highest utility (Algorithm 1, lines 14–15).
5. Apply the offer strategy to pick the "best" proposal found by the GAs in the current round or any of the previous rounds (Algorithm 1, line 16).

Algorithm 1 chooseAction(t, x, z)

Require: m_1, m_2
1: $m_1 \leftarrow$ calculateAspirationValue(t)
2: $m_2 \leftarrow$ calculateMaxDistance(t)
3: **if** $f_1(x) \geq m_1$ **then**
4:    accept(x)
5:    **return**
6: **else**
7:    **if** $f_1(z) \geq m_1$ **then**
8:        propose(z)
9:        **return**
10:    **end if**
11: **end if**
12: newFound $\leftarrow$ globalGeneticAlgorithm()
13: found $\leftarrow$ found $\cup$ newFound
14: newFound $\leftarrow$ localGeneticAlgorithm(x)
15: found $\leftarrow$ found $\cup$ newFound
16: proposeBest(found, m_1, m_2)

---

4 Acceptance Strategy

The acceptance strategy of GANGSTER is given in lines 3–11 of Algorithm 1. It depends on a function of time $m_1(t)$ that we call our *aspiration level*. We do not have space here to explain how it is calculated, but the only important thing to know is that it is a decreasing function of time that represents our agent's willingness to concede.

Let $x \in Agr_n$ denote the last offer proposed by the opponent, and let $z \in Agr_n$ denote the offer with highest utility among all deals made by the opponent in the earlier rounds:

$$\forall z' \in \mathcal{H}_{2 \rightarrow 1}(t) \; : \; f_1(z) \geq f_1(z')$$

If the utility $f_1(x) \geq m_1$, then our agent immediately accept the proposal made by the opponent. If not, then our agent compares its aspiration level with the highest utility offered by the opponent so far, $f_1(z)$. If $f_1(z) \geq m_1$ then α_1 *reproposes* that deal to the opponent. Note that this means that z is a deal that was earlier rejected by our agent (because at that time its aspiration level was higher), but now that the deadline has come closer it has lowered its standards since the risk of failure has become bigger and is now willing to accept it after all. Unfortunately, the alternating offers protocol does not allow an agent to accept a proposal from an earlier round, so it needs to be proposed again. If, on the other hand $f_1(z) < m_1$ it means that it does not consider z good enough (yet), so it will apply the search strategy (Sect. 5) and offer strategy (Sect. 6) to determine a new proposal to make.

Let us now compare this strategy with existing acceptance strategies. In [1] a study was made of several acceptance strategies. The most common acceptance strategy they identified is named $AC_{prev}(1, 0)$. In that strategy the agent α_1 compares the utility of the opponent's offer $f_1(x)$ with the utility $f_1(w)$ of the proposal w that agent α_1 would make if it would not accept x.

Our strategy is a variation of that strategy. However, instead of comparing the utilities of two proposals, our agent compares the utility of the opponent's proposal $f_1(x)$ with its aspiration level $m_1(t)$. The reason for applying this strategy is that α_1 can already determine whether or not to accept the opponent's offer x before it has determined its own proposal w. This has two advantages: firstly, this may save some time because determining the next proposal w can be time consuming. Secondly, it may not always be possible to find a deal w for which the utility is higher than the aspiration level. For example, the agent may only be able to find a deal w with utility $f_1(w) = m_1 - 0.1$. If it would apply $AC_{prev}(1, 0)$ it would not accept the opponent's proposal x, even though x yields a utility value higher than our agent's aspiration level. That would be suboptimal, since the aspiration level is by definition the amount of utility that the agent considers high enough to accept.

Another improvement with respect to the $AC_{prev}(1, 0)$ strategy is that we introduce the concept of *reproposing* (Algorithm 1, lines 6–11). After all, the fact that our agent rejected an earlier proposal from the opponent does not have to mean it would never accept it. The great advantage of reproposing, is that we know that the opponent has already proposed it, and therefore it is very likely that he will accept it. Moreover, the fact that it was already proposed to α_1 means that our agent does not have to search for a new proposal, it already has z readily available in its memory.

5   Search Strategy

We will now explain the Genetic Algorithms (GA) that our agent applies to find good deals to propose. For more background information about Genetic Algorithms we refer to, for example, [4, 11]. To apply a Genetic Algorithm one needs to model the elements of the search space as vectors, called *chromosomes*. Luckily, in the ANAC-domain the possible deals are already given as vectors, so we do not have to put any effort in this modeling. The chromosomes are simply the vectors x as defined in Sect. 2.2. Our GA consists of the following steps:

1. **Initial population**: Randomly pick 120 vectors from Agr_n. This is the initial population.
2. **Selection**: Pick the 10 vectors with highest utility from the population. These are the 'survivors'.
3. **Mutation**: Pick another random vector from Agr_n and add it to the survivors.
4. **Cross-over**: For each pair (v, w) of these 11 survivors, create two new vectors v' and w' (so in total we create 110 new vectors).
5. **New population**: The new population now consists of the 110 new vectors from step 4, plus the 10 survivors from step 2. Go back to step 2, and repeat until convergence, or until we have iterated 10 times.

After a number of iterations the population may contain the same vector more than once. Therefore, when we pick the 10 vectors with highest utility in the selection step, we mean the 10 best *unique* vectors. In other words: we first remove any duplicates from the population and then pick the 10 best vectors. It may happen however that the population has evolved so quickly that no more new unique vectors are created by cross-over. In that case we say it has converged, and the GA is stopped.

5.1 Cross-over

A common way for a GA to apply cross-over, is to cut two vectors both in two halves, and then gluing the first half of one vector to the second half of the other vector and vice versa. We have however opted for a different kind of cross-over, in which random vector-entries are swapped.

Suppose we have two vectors $v, w \in Agr_n$. The cross-over mechanism will output two vectors v' and w' as follows. It first generates a random vector r of dimension n, where each entry r_i has the value 0 with probability 50 % or the value 1 with probability 50 %. Then, given the vectors v, w and r, the vectors v' and w' are defined according to:

$$\text{if } r_i = 0 \text{ then } v'_i = v_i \text{ and } w'_i = w_i$$

$$\text{if } r_i = 1 \text{ then } v'_i = w_i \text{ and } w'_i = v_i$$

The reason that we have chosen for this type of cross-over, is that (as far as the participants can know) there is no relation between the vector entries in the domain. A constraint may for example involve the 3rd and the 7th entry, and there is no reason to assume that consecutive entries are stronger related than non-consecutive entries. This is reflected in our cross-over mechanism by the fact that it is symmetric under any permutation of the entries, whereas the regular cross-over mechanism has a strong bias towards the survival of consecutive sequences of values.

5.2 Global Search Versus Local Search

As we can see in Algorithm 1, in each turn our agent applies two GAs. The first is called the *global* GA, and the second one we call the *local* GA. The difference is that in the global GA it picks vectors randomly from anywhere in the agreement space, while in the local GA it only picks vectors that are close to the last proposal made by the opponent. Specifically: there is a decreasing time-dependent function $m_2(t)$ and our agent only picks vectors for which the Manhattan distance to the last proposal made by the opponent is smaller than $m_2(t)$. The idea is that on one hand α_1 wants to maximize its own utility f_1 and therefore searches for good deals anywhere in the space, but on the other hand also needs to find proposals that are good for the opponent so it applies a local GA to find good deals that are similar to the proposals made by the opponent.

6 Offer Strategy

In lines 12–15 of Algorithm 1 we see that the vectors returned by the GAs are added to a set of potential proposals. After that, it is the task of the offer strategy to determine which of those potential proposals should be proposed to the opponent (Algorithm 1, line 16). In this section we use the notation $d(x, y)$ to denote the Manhattan distance between vectors x and y:

$$d(x, y) = \sum_{i=1}^{n} |x_i - y_i|$$

For each vector x in the set of potential proposals the agent determines three properties, called *utility*, *distance*, and *diversity*, that will determine which deal is the best to propose. The first of these properties, utility, is the most obvious: the higher our agent's utility $f_1(x)$, the better the deal.

Definition 3 The **distance** $dist_t(x)$ of a vector $x \in Agr_n$ at time t, is the lowest Manhattan distance between x and any proposal previously made by the opponent:

$$dist_t(x) = \min\{d(x, y) \mid y \in \mathcal{H}_{2 \to 1}(t)\}$$

The idea is that, since our agent cannot know $f_2(x)$ it uses $dist_t(x)$ as a measure for the opponent's utility instead. If $dist_t(x)$ is low, then there is a high probability that $f_2(x)$ is high. Therefore, our strategy prefers to propose deals with low distance.

Definition 4 The **diversity** $div_t(x)$ of a potential proposal $x \in Agr_n$ at time t is the shortest Manhattan distance between x and any of the proposals previously made by our agent:

$$div_t(x) = \min\{d(x, y) \mid y \in \mathcal{H}_{1 \to 2}(t)\}$$

Our offer strategy prefers proposing deals with high diversity because this has two advantages:

- If α_2 rejected proposal v, and the vector w is close to v (i.e. $div_t(w)$ is low), then it is likely that α_2 will also reject w. So our agent should avoid proposing deals that are similar to earlier rejected deals.
- By proposing more diverse offers, α_1 gives the opponent more information about its utility function f_1, making it more easy for α_2 to find proposals that are profitable to α_1.

Let us clarify this a bit more. Imagine that α_1's utility function has one very high peak. That is: there is a small area inside Agr_n where f_1 is very high. Then α_1 would be inclined to only make proposals from that area. However, if the opponent's utility f_2 is very low in that same area this will be an unsuccessful strategy. Now, suppose that there are a number of other areas of Agr_n where f_1 is less high, but still high enough to be proposed. Then by giving priority to deals with high diversity, we make sure that α_1 also makes proposals around those alternative peaks, hence increasing the probability that for some of these proposals the opponent utility f_2 will also be high. Secondly, in this way α_1 reveals to α_2 the locations of the alternative peaks, which also makes it easier for α_2 to find deals that are profitable to α_1.

We will now explain how utility, distance and diversity are used to determine which proposal to make. This strategy depends on two time-dependent functions: the aspiration level $m_1(t)$ and the maximum distance: $m_2(t)$. Let X denote the set of potential proposals found by the local and global GAs. Then we define the subset $Y \subseteq X$ as:

$$Y = \{y \in X \mid f_1(y) \geq m_1(t) \wedge dist_t(y) \leq m_2(t)\}$$

The deal that α_1 will propose next is then defined as the element $y^* \in Y$ with highest diversity:

$$\forall y \in Y : div_t(y^*) \geq div_t(y)$$

We see here that m_1 acts as a minimum amount of utility α_1 requires for itself, while m_2 acts as a minimum amount of utility that α_1 considers necessary to offer to α_2 (recall that low distance represents high opponent utility). We do not have space to explain how m_1 and m_2 are calculated, but the important thing to know is that both are decreasing functions of time. This means that as time passes, α_1 requires less and less utility for itself, while it forces itself to offer more and more utility to α_2. After all, the closer it gets to the deadline, the more desperate the agent will get to make a proposal that gets accepted by the opponent. If there is more than one potential proposal for which the utility is high enough and the distance is low enough then α_1 picks the one with highest diversity.

7 Motivation for Using Manhattan Distance

Let us now explain why we have chosen to use the Manhattan distance in our definitions. The idea is that we use distance to measure the difference between the utility values of two deals. The closer two deals x and y are, the more likely that $f_i(x)$ is close to $f_i(y)$. Indeed, the utility of a deal is determined by the constraints that it satisfies and if two deals are close to each other then they are likely to satisfy the same constraints. The question however, is which distance measure best reflects the similarity in utility values.

Let c be a constraint that is satisfied by x, that is: $x \in s_c$. Now for each entry x_j the constraint defines two integers a_j and b_j which are unknown. This means that if we increase or decrease x_j by 1, there is a probability that x will no longer satisfy the constraint as x_j may 'drop out' of the interval $[a_j, b_j]$. If x_j is in the interval $[a_j, b_j]$ then we denote the probability that $x_j + 1$ or $x_j - 1$ is not, by p:

$$P(x_j \pm 1 \notin [a_j, b_j] \mid x_j \in [a_j, b_j]) = p$$

Since we have no reason to assume that any entry $j \in [1, n]$ is different from any other entry, we can assume that p is equal for each entry j. Then the probability of dropping out of the constraint after making k steps is p^k, *independent of the directions* of these steps. Specifically: it does not matter whether we take two steps in the $j = 1$ direction or two steps in the $j = 2$ direction, or one step in the $j = 1$ direction and one step in the $j = 2$ direction. In other words, the probability that y satisfies c equals $p^{d(x,y)}$ where d is the Manhattan distance.

Note that this is a direct consequence of the fact that constraints are defined by *rectangular* subspaces. If they had been defined by spherical subspaces for example, then the same reasoning would have lead us to use Euclidean distance.

8 Conclusions

The ANAC'14 competition had two categories: the individual category, in which the agents where ranked according to the individual utility they obtained, and the social category in which agents were ranked by the social utility, which is the sum of the agent's own utility and its opponent's utility. Gangster ranked 3rd place int the individual category, and 2nd place in the social category, among more than 20 participants. We conclude that our agent is a good negotiator and that Genetic Algorithms are a good search technique for the given domain.

However, although the negotiation domains were very large, we think that this may have had only very little influence on the success of the negotiators. The reason for this belief is that when testing our GA on the largest test domain it was often able to find deals with $f_1(x) > 0.95$ in less than 70 ms. This is very short in comparison to the total amount of 180 s available. Therefore, we think the success of a participant depended more on its bargaining strategy than on its search algorithm. This is a pity, because the search was supposed to be the distinguishing property of this year's competition with respect to other years. We would therefore be very interested to know what the results would have been if the deadlines had been much shorter.

Acknowledgments This work was supported by the Agreement Technologies CONSOLIDER project, contract CSD2007-0022 and INGENIO 2010 and CHIST-ERA project ACE and EU project 318770 PRAISE.

References

1. Baarslag, T., Hindriks, K., Jonker, C.: Acceptance conditions in automated negotiation. Complex Automated Negotiations: Theories. Models, and Software Competitions, pp. 95–111. Springer, Heidelberg (2013)
2. de Jonge, D., Sierra, C.: NB^3: a multilateral negotiation algorithm for large non-linear agreement spaces with limited time. J. Auton. Agents Multi-Agent Syst. (2015). In press
3. de Jonge, D., Sierra, C.: Automated negotiation for package delivery. In: 2012 IEEE Sixth International Conference on Self-Adaptive and Self-Organizing Systems Workshops (SASOW), pp. 83–88 (2012). doi:10.1109/SASOW.2012.23
4. Falkenauer, E.: Genetic Algorithms and Grouping Problems. Wiley, NY (1998)
5. Faratin, P., Sierra, C., Jennings, N.R.: Negotiation decision functions for autonomous agents. Robot. Auton. Syst. **24**(3–4), 159–182 (1998). doi:10.1016/S0921-8890(98)00029-3. Multi-Agent Rationality
6. Faratin, P., Sierra, C., Jennings, N.R.: Using similarity criteria to make negotiation trade-offs. In: International Conference on Multi-Agent Systems, ICMAS'00, pp. 119–126 (2000)
7. Marsa-Maestre, I., Lopez-Carmona, M.A., Velasco, J.R., de la Hoz, E.: Effective bidding and deal identification for negotiations in highly nonlinear scenarios. In: Proceedings of The 8th International Conference on Autonomous Agents and Multiagent Systems—Volume 2, AAMAS '09, pp. 1057–1064. International Foundation for Autonomous Agents and Multiagent Systems, Richland, SC (2009). http://dl.acm.org/citation.cfm?id=1558109.1558160
8. Marsa-Maestre, I., Lopez-Carmona, M.A., Velasco, J.R., Ito, T., Klein, M., Fujita, K.: Balancing utility and deal probability for auction-based negotiations in highly nonlinear utility spaces. In: Proceedings of the 21st International Jont Conference on Artifical Intelligence. IJCAI'09, pp. 214–219. Morgan Kaufmann Publishers Inc., San Francisco (2009)
9. Nash, J.: The bargaining problem. Econometrica **18**, 155–162 (1950)
10. Rosenschein, J.S., Zlotkin, G.: Rules of Encounter. The MIT Press, Cambridge (1994)
11. Schmitt, L.M.: Theory of genetic algorithms. Theoret. Comput. Sci. **259**(12), 1–61 (2001). doi:10.1016/S0304-3975(00)00406-0

AgentM

Makoto Niimi and Takayuki Ito

Abstract The Automated Negotiation Agents Competition (ANAC2014) was organized [1]. Automated agents can reduce efforts required of people during negotiations and help people for agreeing in negotiations. Therefore, development of automated agents is one of important issues. We developed AgentM based on a heuristic for the ANAC2014 competition. AgentM has three characteristics described as following:

- AgentM is basically a compromiser agent.
- The purpose of AgentM's bidding strategy is to improve own utility and opponent's utility.
- AgentM's accepted strategy is to accept opponent's bid when it is better than the worst own bid.

In this paper, we show AgentM how to compromise, how to improve utility of each other and how to bid.

1 Introduction

In the fifth international competition (ANAC2014), researchers proposed agents that had various strategies. It is likely that the strategies of an agent can be applied to real-life negotiation problems. Therefore, development of automated agents is one of important issues.

We developed a negotiation agent (AgentM). In this paper, we show AgentM's strategy. Section 2, we describe how our agent compromises. Section 3, we describe how our agent bids. Section 4, we describe when our agent accepts opponent's bid. Finally, we conclude our agent.

M. Niimi (✉)
Department of Computer Science, Nagoya Institute of Technology, Nagoya, Japan
e-mail: niimi.makoto@itolab.nitech.ac.jp

T. Ito
Master of Techno-Business Administration, Nagoya Institute of Technology, Nagoya, Japan
e-mail: ito.takayuki@nitech.ac.jp

© Springer International Publishing Switzerland 2016
N. Fukuta et al. (eds.), *Recent Advances in Agent-based Complex Automated Negotiation*, Studies in Computational Intelligence 638,
DOI 10.1007/978-3-319-30307-9_15

2 Compromising Technique

AgentM compromises using the following factors:

- Time of negotiation
- Distance between an opponent's best and worst bid

We show compromising function which is calculated by Eq. 1.

$$CUV = 1 - (0.1t + d^2) \tag{1}$$

Here, t and d denote a time of negotiation and a distance between an opponent's best bid and worst bid. t and d value's range is between $0 \le t, d \le 1$, because a negotiation's domain of ANAC is normalization.

These are the two reasons why we uses time for compromising.

First, a negotiation has a discount factor. If time of a negotiation is longer, the utility decreases each other. It means agents lose a amount of utility. We design agent is to increases utility according to an early success of a negotiation.

Second, agents succeed for negotiations. When agents fail to negotiate, agents can get only a reservation value. An reservation value is the minimum utility. We design our agent that avoid failing an negotiation.

Next, we describe the reason why uses a distance between an opponent's best bid and worst bid. In ANAC, agents cannot know opponent agent utility spaces. Therefore, our agent uses distance as a guideline that is an opponent agent compromises whether or not. Our agent adjust opponent agent's compromising strategy.

When opponent agent is a early compromised agent, our agent try to succeed earlier for maximizing social welfare. When our agent can succeed earlier, social welfare is not affected by a discount factor.

When opponent agent is a late compromised agent, our agent try to delay agreement for not exploiting our agent's utility. Our agent also exploits opponent agent's compromise.

3 Bidding Strategy

AgentM's bidding strategy is to combination of three bids:

- Best bid
- Best offered bid
- Frequency bid

We explain each of three bids.

Best bid is the highest utility bid for our agent. Our agent searches the best bid using Simulated Annealing (SA). This is attributed to negotiation's domains. In ANAC2014, negotiation's domains are non-liner utility functions. It is difficult for agents to find the best bid. Because issues combination is very huge. For example, 10 issues domain has 10^{10} combinations. It is hard for agents to try all combinations in real time. SA enables agents search a best bid in real time.

Best offered bid is the highest utility bid in opponent's bid until now. Our agent update the best offered bid when opponent agent bid is higher than the best offered bid. Using the best offered bid is due to follow opponent bid. We assume that social welfare of our agent's bid is better to model opponent bid in heuristics.

Frequency bid is tentative bid of the opponent's best bid. Our agent resolve bid into and count up the number of issue values. Frequency bid can help what opponent agent needs. A detailed description of Frequency bid depicts.

Our agent maintains a matrix that counts the number of bids to see the frequency of each issue value for each issue. This matrix is called Issue value frequency matrix: the vertical line shows the issue numbers and the horizontal one shows the issue values. We show an example of Fig. 1 how to update the matrix when our agent receives opponent bids.

If in first opponent's bid the value at issue 0 is 4, the count at value 4 of issue 0 of the matrix increments. Other value of the matrix increments similarly.

After our agent updated the matrix, a frequency bid is generated. We show an example of Fig. 2 how to generate a frequency bid.

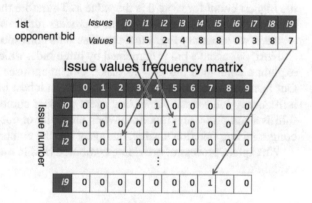

Fig. 1 Example: count up issue values

Fig. 2 Example: generating frequency bid

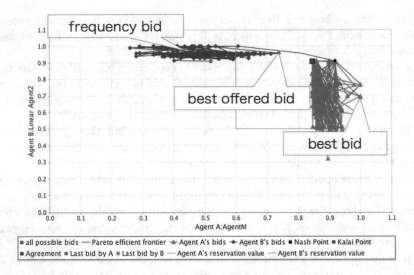

Fig. 3 Three bids correspond to positions

Frequency bid gather the issue value with the most counts. In Fig. 2, for example, the highest count for issue 0 is the value 1. Therefore the frequency bid has value 1 for issue 0. Our agent repeats the same process for the rest of the issues.

Here, we summary Fig. 3 that the three bids correspond to positions.

Next, our agent's bid is combined by three bids, which calls a combined bid. We explain a combined bid how to make. Our agent uses the best bid as a base bid. Our agent replace a best bid's issue value with a best offered bid's issue value and a frequency bid's issue value. Our agent product combined bid randomly while it fulfills a compromising function. When an utility of our agent's bid is higher than a compromising function's value, our agent bid the combined bid.

We explain process of providing combined bids in more details. We use Fig. 4 for explaining.

Fig. 4 Example: process of providing combined bid

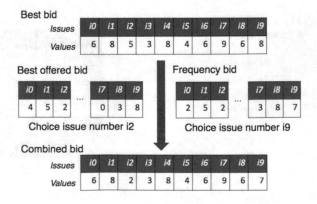

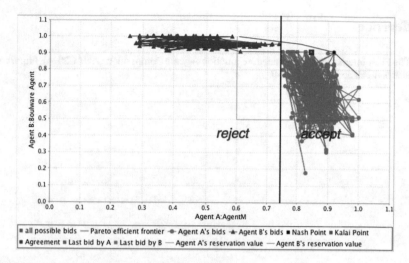

Fig. 5 Example: AgentM rejects or accepts

The best bid has the value 6 at issue 0, the value 8 at issue 1 and so on. Our agent randomly selects issue number 2 from the best offer bid and issue number 9 from the frequency bid. Our agent substitutes the value at issue number 2 by the value from best offered bid at issue number 2 and the value at issue number 9 by the value from frequency bid at issue number 9. As a result, a combined bid is the Fig. 4 at the bottom.

4 Accepted Strategy

We show AgentM's accepted strategy. Our agent's accepted strategy is simple. Our agent accepts that opponent bid when it is higher than worst of our bids until now. Figure 5 is an example of accepting strategy (Fig. 5).

In Fig. 5, the vertical line is a border line whether reject or accept.

5 Conclusion

We show a compromisation technique, bidding strategy, and an acceptance strategy. AgentM's compromisation function consists of time of negotiation and a distance between an opponent's best and worst bid. AgentM's bidding strategy uses a combined bid. A combined bid consists of a best bid, best offered bid, and frequency bid. AgentM's accepted strategy is to accept that opponent bid when it is higher than worst of our bids until now.

Reference

1. The Fifth International Automated Negotiation Agents Competition (ANAC2014). http://www.itolab.nitech.ac.jp/ANAC2014/

k-GAgent: Negotiating Agents Considering Interdependencies Between Issues

Shinji Kakimoto and Katsuhide Fujita

Abstract Bilateral multi-issue closed negotiation is an important class for real-life negotiations. Usually, negotiation problems have constraints such as a complex and unknown opponent's utility in real time, or time discounting. Recently, the attention of this study has focused on the nonlinear utility functions. In nonlinear utility functions, most of the negotiation strategies for linear utility functions can't adopt to the scenarios of nonlinear utility functions. In this chapter, we propose the estimating method for the pareto frontier based on the opponent's bids. In the proposed method, the opponent's bid is divided into small elements considering the combinations between issues, and counted the number of the opponent's proposing to estimated the opponent's utility function. In addition, the genetic algorithm is employed to the proposed method to search the pareto optimal bids based on the opponent's estimated utility and own utility. Experimental results demonstrate that our proposed method considering the interdependency between issues can search the pareto optimal bids.

1 Introduction

In this chapter, we proposed the estimating method for finding the pareto frontier based on the opponent's bids. In the proposed method, the opponent's bid is divided into small ones considering the combinations between issues, and counted the number of the opponent's proposing to estimated the opponent's utility function. In addition, *SPEA2* [2] which is based on the genetic algorithm for multiple objective optimization is employed to our proposed method to search the pareto optimal bids based on the opponent's estimated utility and own utility. In the experiments, we evaluated the

S. Kakimoto (✉) · K. Fujita
Faculty of Engineering, Tokyo University of Agriculture and Technology,
Tokyo 184-8588, Japan
e-mail: kakimoto@katfuji.lab.tuat.ac.jp

K. Fujita
e-mail: katfuji@cc.tuat.ac.jp

© Springer International Publishing Switzerland 2016   241
N. Fukuta et al. (eds.), *Recent Advances in Agent-based Complex
Automated Negotiation*, Studies in Computational Intelligence 638,
DOI 10.1007/978-3-319-30307-9_16

quality of the pareto frontier in our approach measured by the size of the dominant area [3]. The experimental results demonstrate that our proposed method considering the interdependency between issues can search the pareto optimal bids.

The remainder of the chapter is organized as follows. First, we describe related works. Second, we show the negotiation environments and nonlinear utility functions. Third, we describe the *SPEA2* which can search the pareto optimal bids effectively, and propose the novel method for estimating the opponent's utility function. Then, we demonstrate the experimental analysis of finding the pareto optimal bids. Finally, we present our conclusions.

2 *SPEA2* for Finding Pareto Frontier

SPEA2 (*Strength Pareto Evolutionary Algorithm 2*) is the generic algorithm for finding the pareto frontier in multi-objective optimization proposed by Zitzler [2]. The main advantages of *SPEA2* are an improved fitness assignment scheme, which takes for each individual into account how many individuals it dominates and it is dominated by. The algorithm of *SPEA2* is as follows:

Input: N (population size), $\overline{N}$ (archive size), T (maximum number of generations)
Output: A (nondominated set)
Step1:   Initialization: Generate an initial population P_0 and create the empty archive (external set) $\overline{P_0}$. Set $t = 0$.
Step2:   Fitness assignment: Calculate fitness values of individuals in P_t and $\overline{P_t}$.
Step3:   Environmental selection: Copy all nondominated individuals in P_t and $\overline{P_t}$ to $\overline{P_{t+1}}$. If size of $\overline{P_{t+1}}$ exceeds $\overline{N}$ then reduce $\overline{P_{t+1}}$ by means of the truncation operator, otherwise if size of $\overline{P_{t+1}}$ is less than $\overline{N}$ then fill $\overline{P_{t+1}}$ with dominated individuals in P_t and $\overline{P_t}$.
Step4:   Termination: If $t \geq T$ or another stopping criterion is satisfied then set A to the set of decision vectors represented by the nondominated individuals in $\overline{P_{t+1}}$. Stop.
Step5:   Mating selection: Perform binary tournament selection with replacement on $\overline{P_{t+1}}$ in order to fill the mating pool.

The fitness assignment of *SPEA2* is as follows. First, the number of individuals $S(i)$ dominated the individual i is calculated. $R(i)$ (raw fitness of the individual i) is the sum of the $S(j)$ with the individual j dominates the individual i. $R(i) = 0$ corresponds to a non-dominated individual, while a high $R(i)$ value means that i is dominated by many individuals. The $\sqrt{N + \overline{N}}$-th element gives the distance sought, denoted as σ_i. The density $D(i)$ is defined by $D(i) = \frac{1}{\sigma_i + 2}$. The fitness assignment $F(i)$ is defined by $F(i) = R(i) + D(i)$.

In addition, the truncation method, which is invoked when the nondominated front exceeds the archive limit, has been replaced by an alternative truncation method which has similar features but does not loose boundary points.

In *SPEA2*, the individual means the proposed bid, and the genetic locus mess the issue. The values of each issue means the genes. The crossover is uses as the uniform cross-over.

3 Estimating Opponent's Utility Space Considering Interdependency Between Issues

The proposed method estimates the opponent's utility based on its statistical information by dividing the opponent's bids into some small elements. In the alternative offering protocol, the bids proposed by the opponent many times should be considered as the important and high valued ones. However, it is hard to get the statistical information by counting the all bids simply because the proposed bids are limited in the one-shot negotiation. In addition, we can't estimate effectively in the nonlinear utility functions if the issue interdependencies are ignored. Therefore, we propose the novel method the opponent's bid is divided into small elements considering the combinations between issues, and counted the number of the opponent's proposing to estimated the opponent's utility function.

We assume that the alternative with M issues $\mathbf{s}$ is divided into small parts of the alternative with D elements. $C(D, \mathbf{s})$ defined as a function of outputting the set of combining D elements from the alternative with M elements ($|C(D, \mathbf{s})| = {}_MC_D$). $count(e)$ returns the number of the element e outputted by $C(D, \mathbf{s})$ in the previous proposals. The estimating function of the opponent's utility of $\mathbf{s}$ is defined as follows:

$$U(\mathbf{s}) = \sum_{e \in C(D, \mathbf{s})} count(e) \tag{1}$$

Figure 1 shows an example of dividing the proposed bids. When $\mathbf{s}' = (2, 4, 1)$ is proposed by the opponent and divided some small parts of alternatives in (*the number of divided elements*) $= 2$, $C(2, \mathbf{s}') = \{[i_1 = 2, i_2 = 4], [i_1 = 2, i_3 = 1], [i_2 = 4, i_3 = 1]\}$. We assume that the previous situation was $count([i_1 = 2, i_3 = 1]) = 1$, $count([i_2 = 4, i_3 = 1]) = 4$ before the opponent proposes. In this situation, $count([i_1 = 2, i_2 = 4]) = 1$, $count([i_1 = 2, i_3 = 1]) = 2$, $count([i_2 = 4, i_3 = 1]) = 5$ after this proposal is reflected.

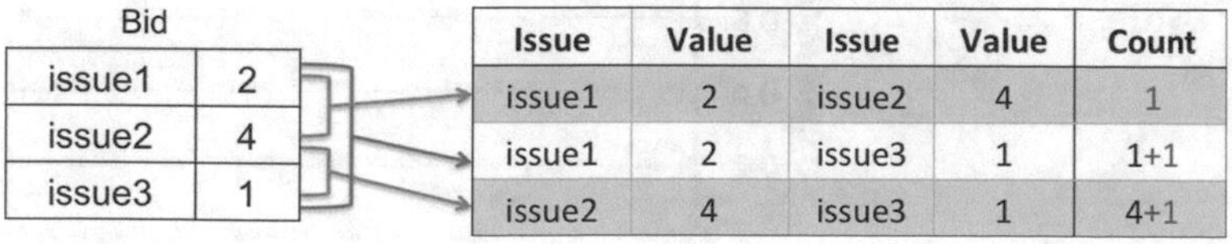

Bid	
issue1	2
issue2	4
issue3	1

Issue	Value	Issue	Value	Count
issue1	2	issue2	4	1
issue1	2	issue3	1	1+1
issue2	4	issue3	1	4+1

Fig. 1 Division of bid and counting up method

After estimating the opponent's utility, our agent can calculate the pareto frontier using our utility function and the opponent's estimated utilities by *SPEA2*.

4   Experimental Results

We conducted several experiments to evaluate our approach. The following parameters were used. The domain for the issue values was [0, 9]. The number of issues is from 10 to 30 at 5 issues intervals, and every experiments are conducted in five different domains. The number of constraints was (*the number of issues*) * 10, and these constraints are related to 3.5 issues on average. The utility of the each agent is decided, randomly.

The opponent proposes in order from the highest utility bid to the lower ones. Figure 2 shows the opponent's proposals in the experiments when the number of opponent's bidding increases. As Fig. 2 showing, the opponent proposes a bid as a gradual compromise by selecting bids with lower utilities for themselves. In the initial phase, most of the negotiating agents wait and see their opponent's bids before judging their opponent's strategy and utilities in the previous ANAC [1]. Negotiating agents usually propose selfish bids to avoid such mistakes as lower utility for their side in the initial phase. After that they propose a bid as a gradual compromise by selecting bids with lower utilities for themselves. Therefore, the opponent's strategy of these experiments is created like the Fig. 2 to evaluate in the previous virtual ANAC.

In the experiments, we evaluated the quality of the pareto frontier in our approach measured by the size of the dominant area [3]. The dominant area increases when the solution set is close to the pareto frontier and exists without any spaces between solutions. This evaluation measure is used the the size of the dominant area in the experiments because it is effective for our approach to find the pareto frontier. The numbers of combinations D are 1, 2 or 3 in the experiments. In *SPEA2*, the number of individuals is 250, the archive size is 250, and 500000 function evaluations. The mutation rate is set as $1/L$. The optimal pareto frontier was found by searching the its

Fig. 2 Example of opponent bids

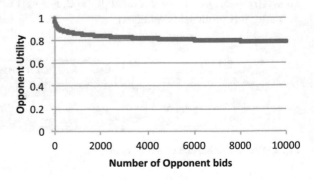

own and the 'opened' opponent's utility spaces by *SPEA2*. Optimality Rate is defined as (*dominant area achieved by each protocol*)/(*optimal dominant area calculated by SPEA2*).

Figure 3 shows the average of the optimality rate in all utility spaces when the number of opponent's bids changes. As the number of opponent's bidding increases, our approach can find the better solutions which is close to the pareto frontier.

Figure 4 shows the average of the optimality rate in all utility spaces when the number of combinations (*D*) changes. Our approach of *D* = 3 can search the better bids which is close to the pareto frontier than that of *D* = 1. Estimating values of our approach of *D* = 1 are calculated by dividing bids with each issue without considering the interdependency between issues. On the other hand, the approach needs to have a power of expressions to the nonlinear utility functions because

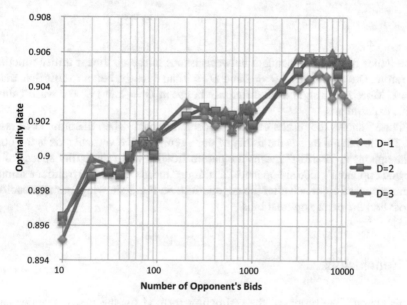

Fig. 3 Dominant area with number of opponent's bids

Fig. 4 Dominant area with combination number

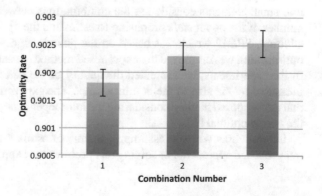

Fig. 5 Searching Pareto
fronts by our proposed
method

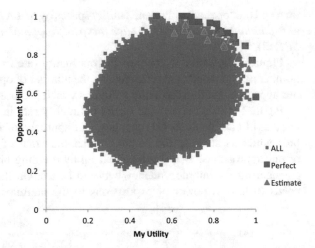

some issues have interdependent between issues in the nonlinear utility functions. Therefore, Our approach of $D = 2$ and $D = 3$ can have the better results than that of $D = 1$. Moreover, a significant difference between $D = 2$ and $D = 3$ wasn't shown in this experiments.

Figure 5 shows the results of our proposed approach when the number of issues is 6. The horizontal axis is the utility of our agent, and the vertical axis is the utility of the opponent. "Perfect" means the pareto frontier when the utility spaces of our side and opponent's side are opened. "Estimate" means the pareto frontier estimated by our proposed approach after the negotiation. As Fig. 5 showing, our approach can almost find the pareto optimal bids.

5 Conclusions

In this chapter, we proposed the estimating method for the pareto frontier based on the opponent's bids. In the proposed method, the opponent's bid was divided into small elements considering the combinations between issues, and counted the number of the opponent's proposing to estimated the opponent's utility function. In addition, *SPEA2* which was based on the genetic algorithm for multiple objective optimization was employed to our proposed method to search the pareto optimal bids. In the experiments, we evaluated the quality of the pareto frontier in our approach measured by the size of the dominant area. The experimental results demonstrated that our proposed method considering the interdependency between issues can search the pareto optimal bids.

Future works will address improvements of estimating the opponent's utility in our proposed approach. For solving this problem, our approach needs to consider the order of opponent's proposals in estimating the opponent's utility. Another important work for our study is to evaluate our approach by having the tournament among our agent and some state-of-the-art agents.

References

1. Aydogan, R., Baarslag, T., Fujita, K., Hindriks, K., Ito, T., Jonker, C: The Fifth International Automated Negotiating Agents Competition (ANAC2014) (2014). http://www.itolab.nitech.ac.jp/ANAC2014/
2. Zitzler, E., Laumanns, M., Thiele, L., Zitzler, E.: Spea 2: Improving the Strength Pareto Evolutionary Algorithm (2001)
3. Zitzler, E., Thiele, L.: Multiobjective evolutionary algorithms: a comparative case study and the strength pareto approach. IEEE Trans. Evol. Comput. 3(4), 257–271 (1999)

A Greedy Coordinate Descent Algorithm for High-Dimensional Nonlinear Negotiation

Bálint Szöllősi-Nagy, David Festen and Marta M. Skarżyńska

Abstract Automated negotiation is a rapidly growing topic in the field of artificial intelligence. Most of the research has been dedicated to linear preferences. Approaching real life negotiations requires more realistic preference profiles to be taken into account. To address this need, nonlinear high-dimensional domains with interdependent issues were introduced. They have however posed a struggle for automated negotiation agents. The difficulty is that fast, linear models can no longer be used in such domains, and there is no time to consider all possible bids in huge spaces. This paper proposes Group2Agent (G2A), an agent that copes with these complex domains and tries to find high Social Welfare outcomes. G2A uses a variant of the Greedy Coordinate Descent (GCD) algorithm, which can scale linearly with the number of issues and is shown to be effective in locating a meaningful middle ground between negotiating parties. Our results show that G2A reaches an average Social Welfare of 1.79, being only 0.03 below the optimal Social Welfare solution and found the optimal solution itself 3 out of 25 times on pre-competition domains. In conclusion, G2A performs among the top ranking agents when it comes to Social Welfare. Furthermore its search algorithm, GCD, scales better than algorithms such as Simulated Annealing used in other agents.

1 Introduction

This article is written in the context of the ANAC competition, which is an annual competition for negotiation agents [1–3]. Most negotiation problems in the real world involve interdependent issues [4], and the current competition, ANAC 2014, aims

B. Szöllősi-Nagy (✉) · D. Festen · M.M. Skarżyńska
Delft University of Technology, Mekelweg 4, Delft, The Netherlands
e-mail: b.szollosi.nagy@gmail.com

D. Festen
e-mail: d.festen@gmail.com

M.M. Skarżyńska
e-mail: skarzynska.marta@gmail.com

© Springer International Publishing Switzerland 2016   249
N. Fukuta et al. (eds.), *Recent Advances in Agent-based Complex
Automated Negotiation*, Studies in Computational Intelligence 638,
DOI 10.1007/978-3-319-30307-9_17

to address the gap between traditional linear techniques, and the challenges posed by a more realistic setting. In the ANAC 2014 competition, the agents competed in a real-time bilateral negotiation using time-discounted nonlinear high-dimensional domains.

In this paper, we propose Group2Agent (G2A), one of the finalists of the ANAC 2014 competition, as an agent to negotiate in complex environments. The focus will be on the inner workings of G2A, how it addresses the challenges of huge contract spaces (10^{30} possible contracts in one domain) and how it handles non-linear utility surfaces arising from issue interdependence [5].

The rest of this paper has the following outline. First, a model-free non-linear search method is introduced. This forms the core of the negotiating agent, it is therefore evaluated against four other algorithms using reference implementations. Subsequently, the overall agent architecture is described, with details of negotiation strategies, opponent modelling and offer acceptance strategy. Where appropriate, design decisions are discussed along the way. An evaluation of the complete negotiation agent is then presented, with results to support its suitability for the competition. The article closes with a summary and directions for future work.

The ANAC competition uses the GENIUS software [6] as the negotiating environment, which is available for download at [7]. The preference profiles, issue domains and computational agents are loaded in the software which then manages the process of negotiation.

2 Description of the Random-Restart Greedy Coordinate Descent Used by Group2Agent

At its core, G2A uses a greedy variation on the Coordinate Descent method (GCD), which itself can be viewed as a variation on the hill climbing algorithm (HC). The main difference with HC is that it looks at neighbors, whereas coordinate descent (CD) does a line search on each issue. In other words, CD maximizes the utility function with respect to one issue while the other issues are held fixed [8], and iterates until a stopping condition.

Coordinate descent is a very simple method that was one of the first optimization approaches proposed and studied [9]. It exists under various names, such as alternate variable method [10], or coordinate relaxation [9]. A clear and concise description is provided in [10], while references for in depth analyses of its behaviour are supplied in [9]. Coordinate descent has not been in the focus of mainstream optimization community, but recently gained popularity due to its successes in tackling high dimensional problems, such as protein loop closure, and truss topology design [9].

An interesting advantage of CD is that it is well suited for parallel computation [11]. The convergence properties of CD are similar to steepest descent, and can be shown to be linear in [11].

The variant of coordinate descent used by G2A uses a greedy strategy to decide which issue to optimize first: It optimizes the issue that gives the best improvement of utility (compared to the initial bid). Other coordinate selection strategies are described in [9].

2.1 Pseudo-code

The algorithm of the greedy coordinate descent is described by the pseudo-code below. Note that the iteration index r is placed in superscript.

> Given parameter bid x^0
> $x^r := x^0$
> **repeat**
> > **for all** $s \in issue\ space$ **do**
> > > In bid x^r, hold fixed all issues other than s
> > > Take issue s in bid x^r, and for each possible option, calculate the utility
> > > Save the resulting 1-D map into L_s
> > > Add L_s to *Local map*
> >
> > Look at *Local map*, select issue and option pair that has the highest utility
> > Store in Q_{issue} and Q_{option}
> > $x^{r+1} := x^r$
> > In bid x^{r+1}, change the issue indexed by Q_{issue} to have the value Q_{option}
> > **if** $x^{r+1} == x^r$ **then**
> > > $stop\ condition := true$
> >
> > $x^r := x^{r+1}$
> **until** *stop condition*
> **return** $x^0, \cdots, x^r$

2.2 Formal Model

The formal model is chosen following [8] but adapted to the greedy coordinate selection strategy, and to the task of maximising over the integer vector domain of the utility function f.

Notation:

$$x \in \mathbb{Z}^p \quad \text{a bid with } p \text{ integer valued issues}$$
$$x^r \in \mathbb{Z}^p \quad \text{a bid at iteration step } r$$
$$x_i^r \in \mathbb{Z}^p \quad \text{the } i\text{-th coordinate of } x, \text{ at iteration step } r$$
$$f : \mathbb{Z}^p \to \mathbb{R} \quad \text{the utility function}$$

Initialization:

Choose any

$$x^0 = (x_1^0, \ldots, x_p^0) \in \mathbb{Z}^p$$

Iteration:

$$r + 1, r \geq 0$$

Given

$$x^r = (x_1^r, \ldots, x_p^r) \in \mathbb{Z}^p$$

Choose an issue

$$s \in \{1, \ldots, p\}$$

Compute new iterate

$$x^{r+1} = (x_1^{r+1}, \ldots, x_p^{r+1}) \in \mathbb{Z}^p$$

Satisfying the following

$$x_s^{r+1} \in \operatorname*{argmax}_{x_s} f(x_1^r, \ldots, x_{s-1}^r, x_s, x_{s+1}^r, \ldots, x_p^r) \tag{1}$$
$$x_j^{r+1} = x_j^r, \forall j \neq s$$

Calculate the above for all issues, and choose the value of s that gives the highest utility.

$$s \in \{1, \ldots, p\}$$

Stopping condition:

$$x^{r+1} = x^r$$

Output of the algorithm:

A path of k steps, from initial bid to local maximum

$$x^t, \text{ where } t \in \{0, \ldots, k\}$$

Satisfying the monotonically increasing utility of each iteration

$$f(x^0) \le f(x^1) \le \cdots \le f(x^k) \tag{2}$$

2.3 Greedy Issue Selection Strategy

The algorithm needs to choose the issue s, which is the next coordinate to evaluate while keeping the others fixed. The classic approach is to select s in a cyclical, round-robin fashion. On the other hand, the greedy issue selection strategy evaluates Eq. 1 for all values of s, and chooses the issue s with the highest utility option. This greedy strategy has a time complexity that is linear w.r.t the number of issues p [9], which means that it scales well into high-dimensional negotiations.

The choice of the initial bid x^0 depends on the negotiation phase. In the opening phase, the initial bids are uniformly sampled from the totality of the bid space. During the middle game phase however, the initial bids are taken from the opponent's bid history, in order to trace a path from an opponent bid towards good bids for the agent.

2.4 Discussion of the Effect of Interdependence

Random restart is used to counterbalance the effect of issue interdependence. Interaction can be seen by looking at the individual issue 'maps' generated by Eq. 1 during line search. Holding all but one issue fixed gives the utility curve of that issue, given the context that consists of the rest of the bid. Without interaction the issues are additive, and the curve will look the same in all contexts. The map is more than a local map, because it covers the full range of an issue, but it is not global, because it is tied to the context where it is made.

Interdependence causes the curve to change shape depending on the context. Figure 1 shows a few examples of context interaction. The utility curves are fitted by their minimum and overlaid. On the 30-dimensional example domain, issue 26 varies smoothly with context, whereas issue 13 is practically reversed depending on the context. Random restart allows the algorithm to discover various contexts, and increase its chance of uncovering useful interactions.

3 Evaluation of the Greedy Coordinate Descent

The random restart greedy coordinate descent was compared to various well-known implementations of nonlinear search methods. The comparison was made in on the nonlinear preference profiles in domain S-1NIKFRT-3, available on the GENIUS website [7]. The profile was the most complex nonlinear preference profile available for

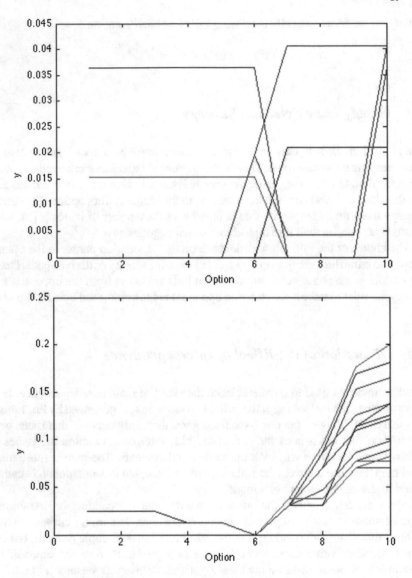

Fig. 1 The effect of interdependence on issue 13 (*above*) and issue 26 (*below*). The utility of each option is shown, with different curves obtained in different contexts

GENIUS before the competition. Results are only shown for the most complex domain, because most methods were too similar for the simpler domains.

For this experiment a custom agent was built that interfaced with Matlab. The utility function of the largest preference profile was exposed in the form of a loss function. Each method was trained on the dataset $N = 25$ times, and the utilities were saved. The default options were chosen and, when the method required it, uniform

Table 1 The best utility found on S-1NIKFRT-3 profile B for various search methods, $N = 25$

Method	Mean	Std	Max utility	Time (min)
Coordinate descent	0.9632	0.0324	1	N/A
Genetic algorithm	0.9525	0.0184	0.9902	7.5777
Pattern Search	0.9442	0.0447	1	2.7909
Simulated annealing	0.8426	0.0268	0.8934	45.589
Global search	0.8339	0.0263	0.8836	6.5906

random starting points were supplied. The total time taken for N runs is listed for Matlab based methods. The GCD used the Java code made for the competition and therefore had no slowdown due to application boundary marshalling which affected the Matlab methods, and this results in the Java code being orders of magnitude faster, and is therefore omitted from the comparison. The results of this experiment are shown in Table 1, and discussed in the next paragraph. The random restart simulated annealing used here is similar to search method implemented by the agent of [12]. The descriptions of all Matlab methods are available on the Mathworks homepage.

Simulated annealing could not find bids with higher utility than 0.90, which made it a less than perfect candidate, along with Global Search. Pattern search and the genetic algorithm were the most promising along with GCD. For more search methods, the interested reader can refer to [13].

The complexity (additive and interdependent) of the utility function was assumed to be representative of the competition domains, therefore GCD was chosen for Group2Agent, based on its ability to find bids above the 0.90 utility level in the most complex domain. An additional advantage of GCD is that it can be used as an anytime algorithm to trace a path between an opponent bid and the closest good bid for us.

4 Description of the Agent

G2A is implemented in the BOA framework [14]. This framework decouples the agent into three parts: (a) A Bidding Strategy or Offering Strategy, which has the responsibility to make an offer, (b) an Opponent Model, which is a model or estimation of the opponent's utility space and (c) an Acceptance Strategy, which decides when to accept the opponent's offer. This separation of concerns makes it easier to reason about individual parts of the application. In this chapter we will discuss each of this components in its respective paragraph.

4.1   Offering Strategy

The Offering strategy aims to propose bids of high joint utility by using a two phase strategy. The first phase offers a selection of best bids from its own profile, and the second phase traces bids between opponent's bid history and own best utility. In the second phase, only opponent bids that have high own utility are taken as initial bids. All the iterations of the GCD algorithm are saved, and the set of resulting bids M are proposed with a time dependent conceder [15].

During the second phase, the GCD is expected to generate bids of high Social Welfare by tracing a path towards the agent's own high utility region. Without knowing the opponent's preferences, one could estimate that a win-win offer lies in M. Equation (2) shows that the steps along the path have a monotonically increasing own utility. In fact, an opponent bid is changed into a good bid for the agent, issue by issue. Figure 2 shows a synthetic example without an opening phase, where the individual iterations of GCD are shown from an opponent bid. The preference profile is the same, S-1NIKFRT-3/B. This traces the path across common ground. In both phases, the contract space was reduced to find high-utility regions. This way of sampling the contract space to find high-utility contract regions was proposed by [4]. In summary, during the middle game, the agent tries to take an opponent offer, and transform it into a good bid for us, with the least amount of transformations it can, with the visibility it has.

The switch from opening phase is made in function of the time discount. Opening game is considered finished, when time becomes larger than $0.25 \cdot D^2$ (where D is the discount factor). The time was chosen to switch around the first quarter of total time, weighed with the discount factor, in this way to promote earlier concession when time is important.

An additional feature of the offering strategy is that the coarse opponent model (OM) is used to filter bids, so that the agent will not make a bid that is much better for the opponent than itself.

4.2   Opponent Model

The OM constructed by taking the marginal distributions of options over each issue in the opponent bid history. This is compared to the expected frequency if each option were chosen uniformly. An option that is proposed more often than expected is counted as a hit, and the number of hits compared to the total number of issues is accumulated. This gives a coarse measure of how similar a bid is to the ones that have passed before.

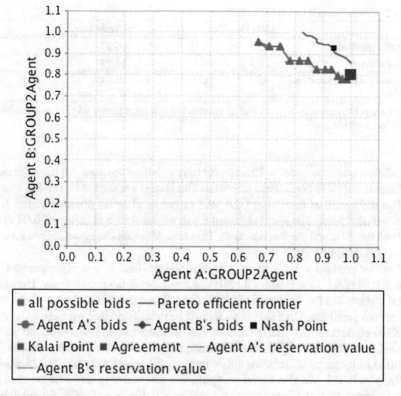

Fig. 2 Tracing the path across common ground in domain S-1NIKFRT-3, starting from an opponent bid towards 1.0 own utility for agent

4.3 Acceptance Strategy

AC Next was used in the competition configuration for its ease of interpretability. AC Next is defined formally in [3], but it can be simplified as follows: If the the utility of the opponent's new bid is higher than the utility of the agent's own last bid, then accept the opponent's offer.

5 Evaluation of the Agent

Besides evaluating the search method, several statistical measurements were also performed on the complete G2A agent. For these tests the S-1NIKFRT-3 domain is used, as this is the most complex domain. Two deadline settings were tested: A deadline of 3 min, as used in the competition and a deadline of 30 s to see how the agent handles shorter deadlines. To get statistically significant information, the

Table 2 Several metrics on the G2A agent

	30 s Deadline	3 min Deadline
Pairwise distance	0.07	0.07
Social welfare	1.78	1.79
Difference with optimal social welfare	0.04	0.03

Tests were performed with $N = 25$ and all numbers in this table have a 95 % confidence interval ± 0.01 the reported number

negotiation was repeated $N = 25$ times for both deadline settings. All the error ranges in this text are 95 % confidence intervals. The results are shown in Table 2.

The main goal in designing G2A was to get as close as possible to the Kalai-Smorodinsky point. The agent obtained a pairwise distance in utility of 0.07 ± 0.01 for both the 30 s, and the 3 m deadline. This indicates that the agent delivers the bids that it was designed for.

Another goal for G2A is to get a high social welfare. In our experiments it gets 1.78 ± 0.01 and 1.79 ± 0.01 for the 30 s and 3 min deadlines respectively. The highest social welfare bid for the domain is 1.82, G2A found this optimal bid 3 times out of 25 for both deadlines. On a side note, it must be mentioned that we were only able to run G2A against itself, which makes the results more biased towards social welfare.

Lastly, we found no significant difference between the 30 s deadline and the 3 min deadline in terms of utilities for either party $t(48) = -1.521$, $p = 0.135$, $t(48) = 0.996$, $p = 0.324$. We also found no significant difference in terms of the used metrics. Distance to Kalai-Smorodinsky $t(48) = -0.404$, $p = 0.608$, Social Welfare $t(48) = 0.165$, $p = 0.870$. Which is an indication that G2A could handle deadlines of 30 s as efficient as the 3 min deadline given in the tournament.

6 Conclusion

Our results show that for the level of complexity exhibited by the nonlinear preference profiles of the competition, performance of GCD is on par with Genetic Algorithm and Pattern Search. Evaluation of the complete G2A agent shows that it is able to find win-win contracts of high Social Welfare in high-dimensional nonlinear negotiations. Because GCD scales linearly with the number of issues, it may be applied when other methods already suffer from the curse of dimensionality (such as Simulated Annealing). The extent of this advantage remains to be quantified. In addition GCD may be implemented as a parallel algorithm to further increase its efficiency.

When applied to negotiation, GCD has a particularly useful property: It allows one to change an opponent bid into a good bid for the agent, with monotonically increasing utility, so as to pass through fair, win-win contracts.

Interdependent issues are handled by applying random restart, which proved to be adequate for the level of complexity of the preference profiles in the competition. However, this approach might hit a limit in case of extreme interdependence. Quantification of this limit is beyond the scope of the article, and remains as future work.

7 Future Work

A practical direction would be to investigate the effect of using more sophisticated acceptance criteria, and more realistic opponent models. A more theoretical direction would compare various methods on preference profiles with extremely high number of issues, in the order of thousands. The curse of dimensionality is expected to affect other methods earlier than coordinate descent, as long as interdependence is not too extreme. An improvement on the method would be to remember the various maps made by each random-restarted GCD particle, and use that to localise regions in the bid space where issues give a higher utility. This improvement would make more of the loss function evaluations than blindly restarting a GCD. Another approach would be to compare random restart GCD with other ways of handling interdependence such as [16].

Acknowledgments We thank Dr. C.M. Jonker and Dr. K.V. Hindriks for introducing us to the field of negotiation and Dr. R. Aydogan and Dr. T. Baarslag for inviting us to the ANAC 2014 competition and supporting us. We also would like to thank S. Rekha for being a team member designing the original linear agent.

References

1. Baarslag, T., Hindriks, K., Jonkers, C., Kraus S., Lin, R.: The first automated negotiating agents competition (ANAC 2010). New Trends in Agent-Based Complex Automated Negotiations, 113–135 (2012)
2. Baarslag, T., Fujita, K., Gerding, E., Hindriks, K., Ito, T., Jennings, N., Jonker, C., Kraus, S., Lin, R., Robu, V., Williams, C.: Evaluating practical negotiating agents: Results and analysis of the 2011 international competition. Artificial Intelligence **198**, 73–103 (2013)
3. Baarslag, T.: What to Bid and When to Stop. Delft University of Technology, Ph.D (2014)
4. Ito, T., Hattori, H., Klein, M.: Multi-issue Negotiation Protocol for Agents: Exploring Nonlinear Utility Spaces. IJCAI **7**, 1347–1352 (2007)
5. Klein, M., Faratin, P., Sayama, H., Bar-Yam, Y.: Negotiating Complex Contracts. Group Decision and Negotiation **12**(2), 111–125 (2003)
6. Lin, R., Kraus, S., Baarslag, T., Tykhonov, D., Hindriks, K., Jonker, C.: GENIUS: AN INTEGRATED ENVIRONMENT FOR SUPPORTING THE DESIGN OF GENERIC AUTOMATED NEGOTIATORS. Computational Intelligence **30**(1), 48–70 (2014)
7. Ii.tudelft.nl, 'Releases – Genius', 2014. [Online]. Available: http://ii.tudelft.nl/genius/?q=article/releases
8. Tseng, P.: Convergence of a block coordinate descent method for nondifferentiable minimization. Journal of optimization theory and applications **109**(3), 475–494 (2001)

9. P. Richtárik and M. Takáč 'Iteration complexity of randomized block-coordinate descent methods for minimizing a composite function', Mathematical Programming, vol. 144, no. 1–2, pp. 1–38, 2012
10. Swann, W.: A survey of non-linear optimization techniques. FEBS Letters **2**, S39–S55 (1969)
11. D. Bertsekas, Nonlinear programming. Belmont, Mass.: Athena Scientific, 1999
12. Hailu, R., Ito, T.: Reducing the complexity of negotiations over interdependent issues. Novel Insights in Agent-based Complex Automated Negotiation, 125–135 (2014)
13. Russell, S., Norvig, P.: Artificial intelligence. Prentice Hall, Englewood Cliffs, N.J. (1995)
14. T. Baarslag, K. Hindriks, M. Hendrikx, A. Dirkzwanger and C. Jonker, 'Decoupling negotiating agents to explore the space of negotiation strategies', Novel Insights in Agent-based Complex Automated Negotiation, pp. 61–83, 2014
15. Faratin, P., Sierra, C., Jennings, N.: Negotiation decision functions for autonomous agents. Robotics and Autonomous Systems **24**(3–4), 159–182 (1998)
16. Hindriks, K., Jonker, C., Tykhonov, D.: Eliminating Interdependencies Between Issues for Multi-issue Negotiation. Cooperative information agents X **4149**, 301–316 (2006)

DoNA—A Domain-Based Negotiation Agent

Eden Shalom Erez and Inon Zuckerman

Abstract Negotiation is an important skill when interacting actors might have misaligned interests. In order to support automated negotiators research the Automated Negotiation Agent Competition (ANAC) was founded to evaluate automated agents in a bilateral negotiation setting across multiple domains. An analysis of various agents' strategies from past competitions show that most of them used an explicit *opponent modeling* component. While it is well known that in repeated interactions, learning the opponent and developing reciprocity become of prominent importance to achieve one's goal, when the interactions with the same partner are *not repeated*, focusing on complex opponent modeling might not be the right approach. With that in mind, we explore a *domain-based approach* in which we form strategies based solely on two domain parameters: the reservation value and the discount factor. Following the presentation of our cognitive model, we present DoNA, a Domain-based Negotiation Agent that exemplifies our approach.

1 Introduction

Negotiation, both an art and science, is often the process used by parties (with possibly misaligned incentives) to reach an agreement regarding cooperation which provides them with mutual benefits. Mathematical models of negotiation, often referred to as bargaining, are ubiquitous in the economic literature [6] and aim at predicting the cooperation details in certain settings. For example such negotiation models are applied to electronic commerce, task allocation, resource management, and supply chains, to name a few [6, 12]. While these models yield important theoretical results

E.S. Erez (✉)
Ariel University, Kiryat Hamada, Ariel, Israel
e-mail: edenerez@gmail.com

I. Zuckerman
Department of Industrial Engineering and Management,
Ariel University, Kiryat Hamada, Ariel, Israel
e-mail: inonzu@ariel.ac.il

© Springer International Publishing Switzerland 2016
N. Fukuta et al. (eds.), *Recent Advances in Agent-based Complex
Automated Negotiation*, Studies in Computational Intelligence 638,
DOI 10.1007/978-3-319-30307-9_18

261

on utility theory, equilibrium points, economic efficiency and managerial insights, the basic assumptions they require in achieving these results (e.g. perfect information and agents' rationality, among others [5, 9]) essentially make these results not transferable nor applicable to real life situations.

To provide researchers with realistic settings to evaluate the performance of a proposed negotiation protocol, the "Automated Negotiation Agent Competition" (ANAC)[1] was founded. This yearly competition allows researchers to employ automated negotiating agents, each carrying out a negotiations strategy, in-order to carry out bilateral negotiation in *non-trivial setting*, and evaluate their relative performance. The evaluation is made in a round-robin fashion where each agent is pitted against each of the other agents in a variety of negotiation scenarios, and the performance scores in each round are aggregated. The ANAC competition is a multi-issue, partial-information, with time discount factor and a reservation value. A review of various agents' strategies from recent competitions show that most of them used an explicit *opponent modeling* component, which is usually boot-started with some default model and is adjusted in real-time during the negotiation session

It is well known that when interactions are repeated, that is when we might negotiate again in the future with the same partner, learning the opponent and developing reciprocity become of prominent importance to achieve one's goal [1, 3]. However, when the interactions (or negotiation in this setting) are a series of one-shot encounters and *are not repeated*, it is somewhat of lesser importance to focus on complex opponent modeling. With that in mind, we set out to design simple automated negotiation strategies for complex negotiation, without any form of explicit opponent modeling component.

To do so, we explore a *domain-based approach* in which we form behavioral strategies based solely on two domain parameters: (1) the reservation value—the payoff that the agent achieves when negotiation concludes without an agreement. (2) the discount factor—the amount of utility lost as time advances. These two parameters are used to divide the class of all possible domains to different regions, in each of which we employ a predefined strategy which is easily motivated by a behavioral intuition. For example, with low discount factor (i.e., our utility depreciates fast with respect to time) and low reservation value we strive to strike an agreement as fast as possible. In contrast, with a high discount factor (i.e., utility does not depreciate much with respect to time) and a high reservation value a favorable strategy is to agree only if the offer is such that we achieve (almost) our maximal possible utility.

Following the presentation of the cognitive model, We present DoNA, an automated negotiation agent that is an implementation instance of the presented domain-based approach. We discuss different implementation parameters that were required to instantiate the agent's algorithm and the heuristic interpretation of the behavioral model.

---

[1]http://ii.tudelft.nl/negotiation/index.php/Automated_Negotiating_Agents_Competition_ (ANAC).

2   The Negotiation Problem

The negotiation environment we consider can be formally described as follows: We studied *bilateral, multi-issue* negotiation setting in which agents negotiate to reach an agreement on a set of several conflicting issues. There is a set of issues, denoted by I. Each issue, $i \in I$, has a finite set of possible values, denoted O_i. An offer, is denoted by $\{o_1, o_2, \ldots, o_{|I|}\}$ where O is the finite set of values for all issues, and $\forall (1 \leq i \leq |I|), o_i \in O_i$.

The negotiations are sensitive to *time*. Time impacts the final utilities of the negotiating parties. Each agent is assigned a *discount factor* which influences its utility as time passes. An agent's final discounted utility is calculated as follows:

$$Discounted\ Utility = Utility * Discount\ Factor^{time}$$

where $time$ and $Discount\ Factor$ are real numbers normalized to the range $[0, 1]$.

The negotiation *protocol* is turn-based, similar to Rubinstein's alternating offers protocol [10]. Each side, in turn, can propose a possible agreement, accept the agreement offered by the opponent at the previous turn, or opt-out of the negotiation.

The negotiation can end either when: (a) the negotiators reach a full agreement, (b) one of the agents opt-out, thus forcing the termination of the negotiation in disagreement, or (c) a pre-specified time limit is reached before an agreement is made. In case (a), each agent achieves a value depending on the agreement reached. In both cases (b) and (c), both agents achieve their reservation values. In all cases, the achieved value is discounted by the cost of time.

Last, we consider environments with *incomplete information*. That is, while both sides have a preference profile, each agent is only aware of its profile and does not have any information about the profile of the other player as well as its discount and reservation values. In the ANAC 2014 competition, for the first time the preference profiles were not specified as a linearly additive utility functions.

3   The Cognitive Model—Playing the Domain

In the negotiation problem presented above, there are three aspects of uncertainty with respect to our opponent: opponent's utility function, opponent's reservation value, and opponent's discount factor. As we **do not** want to explicitly model our opponent, we will not try to estimate the opponent's utility function, but design a strategy based solely on our domains reservation and discount values.

Figure 1 presents a rough division of the set of possible domains based solely on the reservation value (x-axis) and the discount factor (y-axis). Looking at the relationships between the opposing corners of the graph we can see the following. The red diagonal, connecting the Play and the End quadrants, signifies the **temporal flexibility** in the negotiation: In case that the discount factor is low and the reservation

Fig. 1 The quadrants of the domain-based strategic model

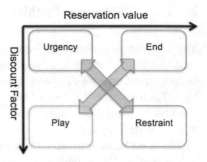

value is high (the "End" quadrant), we wish to spend as little time as possible before either reaching an agreement or opting-out of negotiations, and therefore, this will be a lower bound on the amount of time we should spend on negotiation, implying a strategy where we opt-out of negotiations immediately. On the other edge of the time spectrum is the case where the discount factor is high and the reservation low (the "Play" quadrant), where we want to maximize the time allotted for negotiations in-order to enable an agreement. This can be thought of as an upper bound with respect to how much time to spend negotiating, implying we should allow for a "step-by-step" concessions strategy—e.g., the *Zeuthen* strategy for the Monotonic Concession Protocol [11], which in the perfect information case has been shown to maximize the Nash Product if carried out by both negotiating parties—which should result, eventually in agreement. In between these two bound, some combination of these strategies should be used.

The green diagonal, connecting the Urgency and Restraint quadrants, signifies our **bargaining strength** in the negotiation; On one extreme when we have a high reservation value and a high discount factor (i.e., low cost to the negotiation process) we are in a strong position and can afford being restrained, not making any concessions to achieve agreement, and only achieve agreement in case the other side makes (what may be perceived as very serious) concessions. On the other extreme, with low reservation value and a low discount factor (the "Urgency" quadrant), we find that we are in a weak position, and thus must make many concessions, and fast, to provide the other agent with the proper incentive to accept an agreement. This diagonal, the bargaining strength, should determine the concession strategy applied by the negotiating party.

As our suggested approach does not model the opponent (and to some extent do not even consider the opponent's strategy) we base our analysis regarding the time allocated to the negotiations and our concession stance based on the given reservation value and discount factor. With the above interpretation in mind, the four quadrants provides us with cognitive-dominant behavior strategies.

Play   With a *high* discount factor (and therefore low cost of negotiation time) and a *low* reservation values the parties have a strong incentive to cooperate and achieve an agreement. As the cost of time is low, the parties have enough time to follow some concession protocol, thus participating in a classic interpretation of a negotiation game.

Urgency   With a *low* discount value (that is, high cost of negotiation time) and a *low* reservation value, the parties reason that reaching an agreement is better than disagreement, and due to the high rate of depreciation doing it should be done as quickly as possible. In this case the strategic behavior dictates that a fast application of the concession strategy should be taken. For instance, by increasing the amount of concession steps per round.

Restraint   With *high* discount factor (that is, the cost of negotiation time is low) and a *high* reservation value, the negotiating parties understand that there is room for some benefit from cooperation, albeit very limited. Hence a negotiation session, if one takes place, is restraint, and the parties adopt a stubborn stance (i.e., hold out) and progress very slowly by, for example, reducing the number of concessions per round.

End   With a *low* discount factor and a *high* reservation value, the parties want to end negotiation as fast as possible. This can be achieved by opting-out, or by reaching an agreement if one is agreed upon at the early stages of the negotiation, as time is very costly here.

The above behavioral quadrants span the entire space; While the behaviors are simple and explicit in the end points of the diagonals, any realization of the model should be constructed based on some procedure that takes into consideration the weighted distance of the quadrants strategies from the given reservation and discount values. Accordingly, each diagonal has two points, one min, and one max. These four points located exactly at the corners represent the behaviors which have non-zero weighted values.

4   DoNA—Implementation Details

Based on the behavioral model, we present the automated negotiation agent we have developed. We used the ANAC 2012 data to calibrate our model parameters,[2] e.g. reservation and discount factors thresholds to determine the agent's negotiation strategy, time span to carry out negotiations, the size of concession steps, etc. Following that, we developed a simple way (and somewhat naive) to deal with the new require-

---

[2]Admittedly, we present what we refer to as satisfactory—rather than optimal—parameter values in the sense that the values we used preform well enough, however, we are currently developing the theoretical foundation to calculate optimal parameter values.

ments of the ANAC 2014 competition: non-linear utility function and very large domains.

4.1 The 9 Regions Division

Recalling we focus attention on the values of the discount factor and the reservation value, we discussed the possible behavioral strategies that our agent should be governed by when these values are at one of the 4 corners of the domain. As mentioned between these points a combination of strategies should be used. Therefore, we found it reasonable to partition our state space into a 3-by-3 grid of combination of values, a division which allows us to be somewhat more accurate than a simple 2-by-2 grid yet allows for some differentiation between combinations of intermediate values of discount factors and reservation outcomes.

The following table describes our heuristics we used in the 9 regions of the reservation value (denoted by R) and discount factor (denoted by D) space after normalizing their values to the [0, 1] range. For both parameters, we defined the "low" range as (0, 0.25], the "medium" range as (0.25, 0.75), and the "high" range as [0.75, 1].

The heuristics trying to mimic the behavioral strategy described above are explained in detail below (Fig. 2).

At the high discount factor range (bottom row), we use the **Last** heuristic, which means "wait for the last moment". The rationale is that we have nothing to lose by waiting because the cost of time is low, and we might gain if the opponent decides to concede. The Last heuristic stalls the negotiation (by repeatedly offering the maximum utility attainable and not offering any concessions) until 75% of the time has elapsed. Meanwhile, we estimate the average time, t_r, required for a single round in which each agent makes a single negotiation offer. The "last moment" is defined as t_r before timeout. After 75% of the time has passed, we start making concessions steps in a logarithmic fashion, as follows. Before half the time to timeout has passed (i.e. between 75 and 87.5% of the time has passed), we send all the bids with a utility of at least 0.98 of the maximum attainable value. In the next logarithmic step (i.e. between 87.5 and 93.75% of the time has passed), we send all the bids with a utility of at least 0.96 of the maximum. We proceed in the same way, lowering our threshold in 0.02 utility units each logarithmic time step, until there are 2 rounds to time-out (i.e. the time is, at least, timeout minus $2 \cdot t_r$). At this point, we select between accepting the opponent's offer, opting-out, or offering the bid that is best for us among all bids previously made by the opponent. At the last round (when the time

Fig. 2 The 9 behavioral
regions in DoNA

D ↓ R→	Low	Medium	High
Low	Fast	Fast	End
Medium	Step	Combined	End
High	Last	Last	Last

is timeout minus t_r), we select between accepting the opponent's offer and opting out.

Next, at the high reservation option range (right column) when the discount is medium or low, we use the **End** heuristic, which means "end the negotiation ASAP". The rationale is that we don't have much to gain from negotiation, but if we wait, we loose out on achieving the high reservation value due to the cost of time.

At the low discount factor range (top row) when the reservation value is medium or low, we use the **Fast** heuristic, which means "Make fast concessions to reach an agreement ASAP". In this range, the agent considers all bids in the domain for which its utility is at least as high as the reservation value, and offers them to the opponent in decreasing order of utility, until one of the offers is accepted, or time expires, or the opponent makes a bid with utility higher than the next bid to be offered. The rationale is that the cost of time is high, so we strive to reach an agreement fast even if it means that we make a large concession.

Next, at the low reservation value and medium discount factor range (left-medium cell), we use the **Step** heuristic, which means "make a concession step towards the opponent whenever the opponent makes a step towards you". To implement this heuristic, we keep two lists: a list of all bids in the domain ordered by decreasing order of utility for us, and a list of all bids sent by the opponent. Whenever the opponent sends a new bid (that does not already exist in our list), we send a new bid with the next-highest utility from our list. The rationale is that this cell is an average between the Fast strategy (at the top row), which makes many concession steps without waiting for mutual concessions, and the Last strategy (at the bottom row), which makes only a few concessions and at the final stages of the negotiation.

4.2 The Combined Heuristic

Last, in the middle cell, we use a *Combined* heuristic, which uses an addition parameter, the *fairness equilibrium*. The fairness equilibrium, based on the *Zeuthen strategy* for the Monotonic Concession Protocol (MCP) [11], is the offer that is agreed upon if each of the two agents takes minimal concession steps from its respective maximum utility values until an agreement is reached. This equilibrium is "fair" in the common cultural interpretation as each side takes a similar number of concession steps until the middle point is reached. This occurs when one agent proposes an agreement that is at least as good for the other agent as their proposal:

$$util\_fair = (u_1(x_2) \geq u_1(x_1)) \lor (u_2(x_1) \geq u_2(x_2))$$

In preliminary research, we have conducted an analysis of the domains previously used in the ANAC competitions computed (using Algorithm 1) a cross-domain average $util\_fair$ value of 77 %. This value is almost identical from both players perspectives (max difference <0.03 between the sides). With this parameter we can now

---

Algorithm 1 $util\_fair$ (U_1, U_2)

---

repeat

$\quad x_i \leftarrow \arg\max_{x_i}(U_1(x_i))$

$\quad x_j \leftarrow \arg\max_{x_j}(U_2(x_j))$

$\quad Delete(x_i, x_j)$

until $U_1(x_2) \geq U_1(x_1)$ **or** $U_2(x_1) \geq U_2(x_2)$

---

present our **Combined** heuristic which uses the time dimension in mixing between the strategies mentioned above and does so as follows.

1. Define T_{end} as the time in which we are going to end the negotiation. This time is calculated as the time in which the discount is exactly 0.8, i.e., $Discount$ $Factor^{T_{end}} = 0.8$.
2. During period $[0, 0.25 \cdot T_{end}]$ use the Fast heuristic using a reservation value of $util\_fair \times 120\%$.
3. During period $(0.25 \cdot T_{end}, 0.5 \cdot T_{end}]$ use the Fast heuristic using a reservation value of $util\_fair \times 110\%$.
4. During period $(0.5 \cdot T_{end}, 0.75 \cdot T_{end}]$ use the Step heuristic.
5. During period $(0.75 \cdot T_{end}, T_{end}]$ use the Last heuristic.

The rationale behind the order in the combined heuristic is as follows: first, we define the allotted time for negotiations. During its first part a high utility may be achieved by progressing fast towards an agreement, if we restrict ourselves to high utility offers and note that we want to avoid losses due to the cost of time. If our fast concessions have not helped us achieve an agreement, we slow down and limit the pace of concessions to match those of the opponent using the Step heuristic, which is deemed fairer. At the last period of the negotiation, we have formed a belief that our opponent is not willing to compromise, we allow for the minute chance he may concede but avoid any unnecessary concessions on our part, all the while guaranteeing achieving at least our discounted reservation value.

4.3 Dealing with Non-linear Utility and Large Domains

After developing the agent on the 2012 data and testing it offline on the 2013 data, it was time to put it to the test. As such, we have enrolled DoNA to the 2014 ANAC competition.[3] However, as in previous years, the 2014 edition of the competition issued additional challenges: First, for the first time the utility functions of the domains were extended to include nonlinear utility functions. Second, the organizers have decided to also include very large size domains (e.g. domains of size 10^{50}).

The effects of the new changes meant that the players could query the framework for the utility values of individual bids and get a utility number. However, do to time

---

[3]Competition website: http://www.itolab.nitech.ac.jp/ANAC2014/.

constraints and very large domains it was only possible to get a very small sample of the available bids and their utilities prior to the beginning of the negotiation session (around 100, 000 bids). These changes mean that brute-force methods are not feasible, and previous techniques that were applied to modeling the opponent are also not relevant anymore. Nevertheless, DoNA did not use neither of these methods, so no changes were needed with that respect. However, DoNA was explicitly using the ordering of the bids in its strategy (e.g. when taking step-by-step concessions), this cannot be done explicitly in the new setting. To cope with the new challenges, when we were faced with domains that contained more than one million bids we assumed the utility function has a *normal distribution* on the utility values, and sample 100,000 thousand bids and use them to sort all the bids we have. We also calculate the average and the variance of the sampled bids to set a threshold representing the minimum expected utility (which is a required data for DoNA's strategy) according to $2.4 * \sigma + 1.2 * \mu$ (where σ is the standard deviation, and μ is the average). We did not do anything else concerning to the nonlinear utility functions challenge since it was irrelevant to our model. In smaller domains DoNA acted as described in the original model.

5 Related Work

In this work we are looking at the problem of bilateral, incomplete information, time constrained, multi-issue negotiations. In other words, we have two parties that need to agree on several issues, but are unaware of the preferences of the other side. There is an extensive literature on different versions of the problem and solution approaches, we will cite the most relevant ones to point the differences with our current work. We refer the reader to an excellent review in [7].

Game theoretic models provide a right mathematical tools to analyze and predict the negotiation outcomes. However, the cost of the mathematical rigidity is in the simplifying assumptions that are used. Their well-known equilibriums solutions such as Nash bargaining, or the Kalai-Smorodinsky solutions [5, 9] requires *complete* information. Game theoretic solutions provide many interesting insights on the formal side of the problem, but are usually not practical to apply in reality due to their simplifying assumptions.

Due to the inherent difficulty of the problem and the need for practical solutions, various heuristic models were proposed over the years. Byde et.al. [2] developed *AutONA*, an automated negotiation agent. Their problem domain involves multiple negotiations between buyers and sellers over the price and quantity of a given product. Jonker et.al. [4] created an agent to handle multi-attribute negotiations which involve incomplete information. The *QOAgent* [8] is a domain independent agent that can negotiate with people in environments of finite horizon bilateral negotiations with incomplete information. Recently, the first step in migrating from Dialog-based environments to complete Chat-based interfaces was taken [13].

The usefulness of the different approaches have been recently put to the test in the yearly *ANAC competition* that started in 2010. The competition provides a platform to compare and benchmark different state-of-the-art heuristics developed for automated, complex bilateral negotiation. Nevertheless, none of the previous agents were basing its strategy only on the properties of the domain.

6 Conclusions

The current research direction in constructing complex automated negotiators is through the development of an accurate opponent model. Nevertheless, we conjecture that because negotiations are a one-shot interaction focusing on complex opponent modeling might not be the right path to take. Consequently, we presented DoNA, an automated negotiation agent that utilizes *domain-based approach* to negotiation.

DoNA is built upon a simplified cognitive model that defines the natural-heuristics when considering only the reservation and discount values of the domain. We also provided an interpretation of these strategies in DoNA. While we do not claim that the presented interpretations will provide the optimal results, they were proved to be quite successful as DoNA attained the 2nd place in the individual utility category of the ANAC 2014 competition. This demonstrates that a simple domain-based agent can achieve results comparable to or even better than more complex opponent-based agents, and practically it can be used as a baseline for evaluating the utility of adding opponent-based models to agents

In the future we see several possible directions. First, we conjure that DoNA will be especially successful when negotiating with people because it is built upon cognitive behavioral foundations for negotiation. We feel that human would tend to apply the same reasoning process thus allow it to form an agreement faster. We plan on conducting an experimental session with students shortly. Besides, we are in the process of developing a solid mathematical bargaining model that will be used to tune the parameters to different types of domains ("win-win" vs. "win-loss" vs. "loss-loss"), various sizes of domains (in terms of the number of possible agreements), and others.

References

1. Axelrod, R.M.: The Evolution of Cooperation. Basic Books, New York (1984)
2. Byde, A., Yearworth, M., Chen, K.Y., Bartolini, C.: Aut ONA: A system for automated multiple 1–1 negotiation. In: Proceedings of the 2003 IEEE International Conference on Electronic Commerce (CEC), pp. 59–67 (2003)
3. Cheng, K.L., Zuckerman, I., Nau, D.S., Golbeck, J.: The life game: cognitive strategies for repeated stochastic games. In: SocialCom/PASSAT, pp. 95–102. IEEE (2011)
4. Jonker, C.M., Robu, V., Treur, J.: An agent architecture for multi-attribute negotiation using incomplete preference information. Auton. Agents Multi-Agent Syst. 15(2), 221–252 (2007)

5. Kalai, E., Smorodinsky, M.: Other solutions to Nash's bargaining problem. Econometrica **43**, 513–518 (1975). http://jmvidal.cse.sc.edu/library/kalai75a.pdf
6. Lai, G., Li, C., Sycara, K., Giampapa, J.: Literature review on multi-attribute negotiations. Technical report, Carnegie Mellon University (2004)
7. Lai, G., Sycara, K.: A generic framework for automated multi-attribute negotiation. Group Decis. Negotiat. **18**(2), 169–187 (2009). doi:10.1007/s10726-008-9119-9, http://dx.doi.org/10.1007/s10726-008-9119-9
8. Lin, R., Kraus, S., Wilkenfeld, J., Barry, J.: Negotiating with bounded rational agents in environments with incomplete information using an automated agent. Artif. Intell. **172**(6–7), 823–851 (2008)
9. Nash, J.: The bargaining problem. Econometrica **18**(2), 155–162 (1950)
10. Osborne, M., Rubinstein, A.: A Course in Game Theory. MIT Press, Cambridge (1994)
11. Rosenschein, J.S., Zlotkin, G.: Rules of Encounter—Designing Conventions for Automated Negotiation among Computers. MIT Press, Cambridge (1994)
12. Sim, K.M.: A survey of bargaining models for grid resource allocation. SIGecom Exch. **5**(5), 22–32 (2006). doi:10.1145/1124566.1124570, http://doi.acm.org/10.1145/1124566.1124570
13. Zuckerman, I., Rosenfeld, A., Kraus, S., Segal-Halevi, E.: Towards automated negotiation agents that use chat interfaces. In: The Sixth International Workshop on Agent-based Complex Automated Negotiations. Saint Paul, Minnesota, USA (2013)

WhaleAgent: Hardheaded Strategy and Conceder Strategy Based on the Heuristics

Motoki Sato and Takayuki Ito

Abstract The Automated Negotiation Agents Competition (ANAC2014) was organized. Previous works proposed agents that had various strategies. In this paper, we developed a negotiation agent (WhaleAgent) for ANAC2014. WhaleAgent has two searching bid strategies and two bidding strategies. The searching bid strategies use Simulated Annealing and Hill Climbing. The bidding strategies are Hardheaded strategy and Conceder strategy. We describe these strategies based on the heuristics.

1 Introduction

The Fifth International Automated Negotiating Agents Competition (ANAC2014) was held [1]. At ANAC2014, researchers proposed agents that had various strategies. The strategies of an agent can be applied to real-life negotiation problems.

We developed a negotiation agent (named WhaleAgent) for ANAC2014 that can facilitate on various negotiation problems. Our agent chages behaviors depend on the opponent conditions.

Section 2 describes our agent's searching method using Simulated Annealing and Hill Climbing. Section 3 presents our agent's strategies, Hardheaded strategy and Conceder strategy. The conclusions are given in Sect. 5.

M. Sato (✉) · T. Ito
Nagoya Institute of Technology, Nagoya, Japan
e-mail: sato.motoki@itolab.nitech.ac.jp

T. Ito
e-mail: ito.takayuki@nitech.ac.jp

© Springer International Publishing Switzerland 2016
N. Fukuta et al. (eds.), *Recent Advances in Agent-based Complex
Automated Negotiation*, Studies in Computational Intelligence 638,
DOI 10.1007/978-3-319-30307-9_19

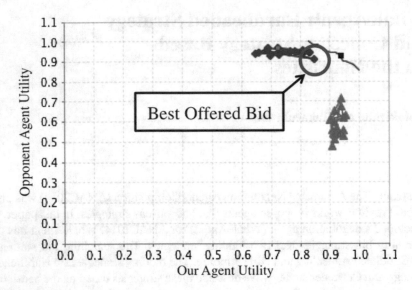

Fig. 1 The best bid for WhaleAgent in the negotiation. WhaleAgent is *green* agent

2 Searching Bids Strategies

In this section, we describe our agent's searching bids strategies.

At ANAC2014, the utility space is a nonlinear utility space. Nonlinear utility spaces have complicated shape since their spaces' constraints contain dependent relationships between the issues. It is difficult to search an agreement candidate for negotiators in nonlinear utility spaces. Therefore, we need to consider a method for searching efficiently on the domain.

Our agent, WhaleAgent uses two searching methods. First searching method is the Hill Climbing. Second searching method is Simulated Annealing.

We express the first searching method using Hill Climbing. Threshold is the value which our agent offered the

1. Our agent chooses the best offered bid from the list of bids the opponent offered before.
2. Our agent uses Hill Climbing with the chosen best offered bid as starting point.
3. If the bid utility is more than Threshold, our agent offers the bid.

We show the best bid in Fig. 1.

Our agent finds good bid with Hill climbing using opponent best offered bid as starting point.

We show the good bid which is searched by Hill climbing in Fig. 2.

When our agent couldn't search the bid which utility is more than Threshold, our agent uses the second searching method based on Simulated Annealing.

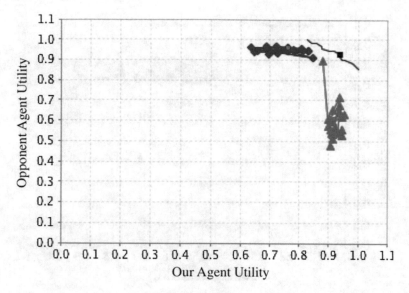

Fig. 2 The good bid which is searched by Hill climbing

We show the second searching method using Simulated Annealing.

1. Our agent uses Simulated Annealing using a random bid as starting point.
2. If the bid utility is more than Threshold, our agent offers the bid.
3. If the bid utility is less than Threshold, our agent searches again.

3   Bidding Strategies

In this section, we describe our agent's bidding strategies. Our agent has two kind of strategies, Hardheaded strategy [2] and Conceder strategy [3].

Our agent changes the strategies depend on the opponent conditions.

3.1   Hardheaded Strategy

We describe Hardheaded strategy. Hardheaded strategy is useful when opponent agent has a conceder strategy.

SA_MAX is the value which our agent found the maximum utility using Simulated Annealing. At the begining of the negotiation, Hardheaded strategy offers the bid which utility is greater than 0.9. At the end of the negotiation, Hardheaded strategy decreases threshold.

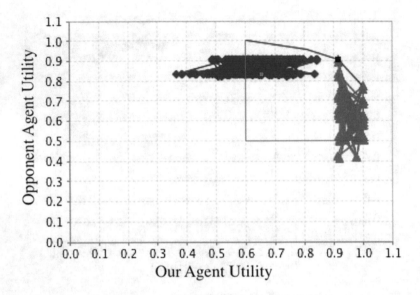

Fig. 3 The Hardheaded strategy

1. Offer bids which utility is greater than $0.9 * SA\_MAX$ ($0.0 <$ time < 0.8)
2. Offer bids which utility is greater than $0.9 * SA\_MAX * 0.99$ ($0.8 <$ time)

We show the agent bidding behavior based on Hardheaded strategy in Fig. 3.

In Fig. 3, our agent's bid represents green points. Our agent offers the bids which utility is greater than 0.9 since SA_MAX is close to 1.0 and threshold is close to 0.9.

3.2 Conceder Strategy

We describe Conceder strategy. If opponent agent offers same bid, our agent changes to conceder behavior. In Fig. 4, the opponent agent offers the same bids. In Fig. 5, opponent agent offers same bids therefore our agent changes to conceder strategy.

Conceder strategy discretes the threshold. SA_MAX is the value which our agent found the maximum utility value using Simulated Annealing. $Best Bid$ is the best offered bid utility value for our agent, which opponent agent offered before.

1. Offer bids which utility is greater than $0.85 * SA\_MAX$ ($0.1 >$ time > 0.0)
2. Offer bids which utility is greater than $0.80 * SA\_MAX$ ($0.5 >$ time > 0.1)
3. Offer bids which utility is greater than $0.70 * SA\_MAX$ ($0.8 >$ time > 0.5)
4. Offer bids which utility is greater than $0.85 * SA\_MAX$ ($0.9 >$ time > 0.8)
5. Offer bids which utility is greater than $Best Bid * 0.99 * SA\_MAX$ ($1.0 >$ time > 0.9)

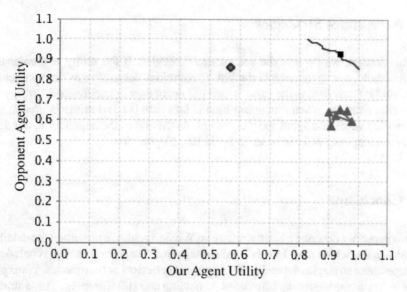

Fig. 4 The opponent agent offers the same bids

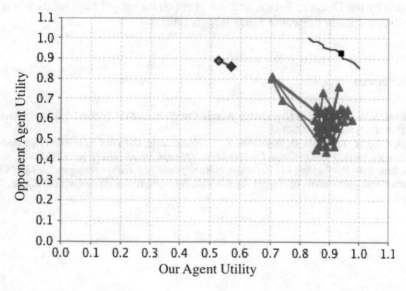

Fig. 5 Conceder strategy

At the begining of the negotiation, conceder strategy decreases the threshold as time passes. Conceder strategy increases the threshold when time is greater than 0.8.

4 Acceptance Strategies

In this section, we express the acceptance strategies. If the utility is greater than the threshold, our agent accepts the bid. Threshold is decreasing as time passes and threshold is equivalent to the value which our agent uses in Hardheaded strategy and Conceder strategy. When remaining time is less than 0.1 and utility is greater than reservation value, our agent accepts the bid. At the end of the negotiation, our agent tends to accept any offer in order to avoid End Negotiation.

5 Conclusion

In this paper, we describe a basic strategy on WhaleAgent. We presented the details of the strategies, which have Hardheaded strategy and Conceder strategy. WhaleAgent changes these strategies depend on the opponent behaviors or time passes. We explain how to search the bids using Simulated Annealing and Hill Climbing. These strategy based on the heuristic can search for the bid with high utility.

Our agent offers the bids regardless of Discount Factor. Therefore our agent affected by the Discount Factor, and our agent could not get high utility when the negotiation domain's discount factor is high value.

References

1. ANAC2014—Automated Negotiating Agents Competition 2014 (ANAC 2014). http://www.itolab.nitech.ac.jp/ANAC2014/
2. van Krimpen, T., Looije, D., Hajizadeh, S.: Hardheaded. In: Complex Automated Negotiations: Theories, Models, and Software Competitions. Springer, Berlin (2013), pp. 223–227
3. Fatima, S.S., Wooldridge, M., Jennings, N.R.: Optimal negotiation strategies for agents with incomplete information. Intelligent Agents VIII. Springer, Berlin Heidelberg (2002), pp. 377–392

Agent YK: An Efficient Estimation of Opponent's Intention with Stepped Limited Concessions

Yoshiaki Kadono

Abstract On my agent, I heavily focused on this efficiency issue to realize a great scalability for the estimation of opponents' preferences. Naturally, to realize it, we should consider some approximation approaches. In my agent, I rather utilized a local-search based approach for searching good bid alternatives with a limited approximate estimation of the opponent's utility space. This heuristic offering approach could also be effective to produce a better outcome as well as a better social welfare when the opponent has a mechanism which tries to sense the degree of intention for a cooperation. However, due to its nature of approximation, it may not always be sensed as a cooperative even when the agent has been tried to be cooperative to the opponent. Especially, on the nearly ending phase of the negotiation, both the agent and the opponent may stick their best compromised proposals together due to this issue. To overcome such a situation, I introduced a special gimmick named "Hotoke no kao mo sandomade", that allows more compromised proposals at most three times. This gimmick was named that the agent will "forgive" the uncooperative reactions of the opponent when the opponent stops such uncooperative reactions after a compromised proposal from the agent was received. The approach is based on my own analysis of the possible behaviors of the agents to be submitted as well as the domains, and of course respect the same proverb that is popular in Japan.

1 Introduction

Estimating opponents' preferences is one of important issues to realize a good strategy for negotiating agents [4]. Even on linear utility spaces, there have been numerous approaches to estimate the opponents' preferences [1, 6]. On non-linear utility spaces, the difficulties as well as the importance on them are more than linear the ones on linear utility spaces [5].

Y. Kadono (✉)
Graduate School of Informatics, Shizuoka University, Johoku,
Hmamamatsu 432-8011, Japan
e-mail: gs14014@s.inf.shizuoka.ac.jp

© Springer International Publishing Switzerland 2016
N. Fukuta et al. (eds.), *Recent Advances in Agent-based Complex
Automated Negotiation*, Studies in Computational Intelligence 638,
DOI 10.1007/978-3-319-30307-9_20

279

In recent ANAC competitions, the *discount factor* has been introduced, which reduces the outcome of the agreement based on the time consumed for the negotiation. In such a situation, taking a long time for estimating the opponent's preference could decrease both the outcomes of the own agent as well as social welfare. In the competition, we could not predict how the used problems called *domains*, which are constructed by the sets of utility spaces to be negotiated. When the number of issues would be greater than several to ten, the computational complexity of calculating estimated opponents' utility spaces could be a serious issue. Therefore, realizing better computational efficiency on such an estimation is important. Furthermore, on linear utility space domains, Baarslag argued that the importance of estimating opponents preferences is less important in some contexts but rather the choice of strategies is important [2]. Actually, at the moment that I have implemented my negotiating agent, I did not know about the Baarslag's paper. However, the shown observations were very similar to my own intuition.

On my agent, I heavily focused on this efficiency issue to realize a great scalability for the estimation of opponents' preferences. Naturally, to realize it, we should consider some approximation approaches. There are a number of known good general approximation algorithms such as genetic algorithms (GA), and simulated annealing (SA). However, in other areas, it is reported that when the complexity of the combinatorial optimization problem is huge, a simple parallel local search approach (so called beam-search) could produce better results [3]. Therefore, in my agent, I rather utilized a local-search based approach for searching good bid alternatives with a limited approximate estimation of the opponent's utility space. In the next section, I will explain some details about my approach.

2 The Approach

2.1 PairBidElements

My agent uses BidElement and PairBidElement in order to approximate acceptable bids. Figure 1 shows the relations among a bid, BidElements and PairBidElements. In this example, we have three issues, shape of frame, shape of wheels, and color of frame. In this case, a bid has some attribute values for all of these attributes. BidElements are a set of those attribute=value pairs for a specific bid. For example, in this case, the three elements, frame=squareShaped, wheel=simpleRoundAlloy, and color=Red, are in the set. PairBidElements are a set filled by the all possible pairs in the BidElements. In this case, frame=squareShaped:wheel=simpleRoundAlloy could be an element in the PairBidElement.

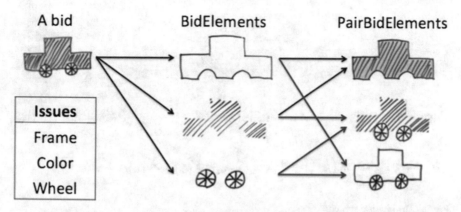

Fig. 1 A Bid, its BidElements, and its PairBidElements

2.2 Bidding Strategy

In agentYK, I just used the correlations among the proposed bids and the opponents' counter offers, as follows:

$$b^* = \arg\max_{b \in B}(U(b) + A(X_{opp}, Y_{opp}, X_{my}, Y_{my}, b))$$

Here, X_{opp} and Y_{opp} are the BidElements and PairBidElements produced from the recent offers received from the opponent. Similarly, X_{my} and Y_{my} are BidElements and PairBidElements produced from the recent offers sent by the agent itself. To calculate b^*, it evaluates how the b bid would be preferable by the opponent, which also considers its own past compromises. Typically, if b is similar to received bids, the function $A(X_{opp}, Y_{opp}, X_{my}, Y_{my}, b)$ returns higher value. The agent offers the bid b^* which maximizes the above value (Fig. 2).

In this approach, when the size of stored bid history becomes huge, the computation for this b^* might take a certain computational time. However, although it depends the design of the function A, it could be implemented in an efficient way since we can limit the similarity computation on the level of PairBidElements, as well as the use of a sharp cut-off function for limiting the number of bid histories utilized in the actual computation.

As mentioned above, this agent totally does not make any direct prediction of the opponent's preference, but tries to propose bids that would be acceptable for the opponent based on co-occurrences in the BidHistories. Furthermore, it also tries to make some concession when the opponent seemed to have responded *cooperative* counter offers.

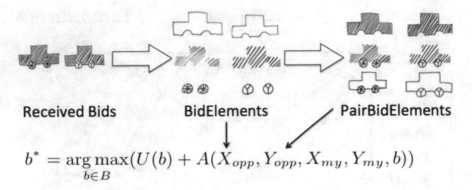

$$b^* = \arg\max_{b \in B}(U(b) + A(X_{opp}, Y_{opp}, X_{my}, Y_{my}, b))$$

Fig. 2 The bidding strategy

2.3 Special Gimmick

The above mentioned heuristic offering approach could also be effective to produce a better outcome as well as a better social welfare when the opponent has a mechanism which tries to sense the degree of intention for a cooperation. However, due to its nature of approximation, it may not always be sensed as a cooperative even when the agent has been tried to be cooperative to the opponent.

Especially, on the nearly ending phase of the negotiation, both the agent and the opponent may stick their best compromised proposals together due to this issue. For example, like the situation shown in Fig. 3, both would not come closer but sticking in a certain position. To overcome such a situation, I introduced a special gimmick

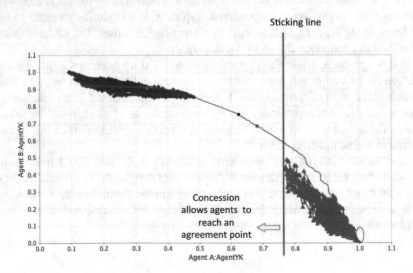

Fig. 3 The "sticking" situation and the use of "SpecialGimmic"

named "hotoke no kao mo sandomade", that allows more compromised proposals at most three times.

The gimmick was named that the agent will "forgive" the uncooperative reactions of the opponent when the opponent stops such uncooperative reactions after a compromised proposal from the agent was received. Based on my own analysis and observation to some previously proposed negotiating agents, such "sticking" might occur multiple times in each negotiation game. I have decided this parameter be three based on my own estimation of the possible behaviors of the agents to be submitted as well as the domains, and of course respect the proverb that is popular in Japan.

3 Comments and Remarks

After the competition, I noticed the submitted domains were rather far simpler than I expected and the used computing devices were so powerful. When the problem space were much more complex, e.g., the number of issues are much larger, my agent will achieve a bit better result. In other words, my idea might have further advantages that were not tested in this year's competition. I hope such advantages might be seen in much deeper analyses and some would use them for more powerful negotiating agents.

Acknowledgments I really appreciate the great help of my supervisor, who supported me to travel to the Paris and attend the conference by his own expense, as well as making the presentation and this brief description of my agent better and readable.

References

1. Baarslag, T., Fujita, K., Gerding, E.H., Hindriks, K.V., Ito, T., Jennings, N.R., Jonker, C.M., Kraus, S., Lin, R., Robu, V., Williams, C.R.: Evaluating practical negotiating agents: results and analysis of the 2011 international competition. Artif. Intell. **198**, 73–103 (2013)
2. Baarslag, T., Hendrikx, M., Hindriks, K., Jonker, C.: Predicting the performance of opponent models in automated negotiation. In: Proceedings of the 2013 IEEE/WIC/ACM International Conference on intelligent Agent Technology(IAT2013), pp. 59–66 (2013)
3. Fukuta, N., Ito, T.: Fine-grained efficient resource allocation using approximated combinatorial auctions-a parallel greedy winner approximation for large-scale problems. Web Intell. Agent Syst. Int. J. **7**(1), 43–63 (2009)
4. Gal, Y.K., Ilany, L.: The fourth automated negotiation competition. In: Next Frontier in Agent-Based Complex Automated Negotiation, pp. 129–136. Springer (2015)
5. Ito, T., Klein, M., Hattori, H.: A multi-issue negotiation protocol among agents with nonlinear utility functions. Multiagent Grid Syst. **4**(1), 67–83 (2008)
6. Koeman, V.J., Boon, K., van den Oever, J.S., Dumitru-Guzu, M., Stanculescu, L.C.: The fawkes agent- the anac2013 negotiation contest winner. In: Next Frontier in Agent-Based Complex Automated Negotiation, pp. 143–151. Springer (2015)

BraveCat: Iterative Deepening Distance-Based Opponent Modeling and Hybrid Bidding in Nonlinear Ultra Large Bilateral Multi Issue Negotiation Domains

Farhad Zafari and Faria Nassiri-Mofakham

Abstract In this study, we propose BraveCat agent, one of the ANAC 2014 finalists. The main challenge of ANAC 2014 was dealing with nonlinear utility scenarios and ultra large-size domains. Since the conventional frequency and Bayesian opponent models cannot be used to model the unknown complex nonlinear utility space or preference profile of the opponent in ultra large domains, we design a new distance based opponent model to estimate the utility of a candidate bid to be sent to the opponent in each round of the negotiation. Moreover, by using iterative deepening search, BraveCat overcomes the limitations imposed by the huge amount of memory needed in the ultra large domains. It also uses a hybrid bidding strategy that combines behaviors of time dependent, random, and imitative strategies.

1 Introduction

In this paper, we elaborate on BraveCat Agent negotiation strategy, which was designed and implemented to take part in ANAC 2014. ANAC is a tournament in which a set of automated negotiating agents compete against each other, in a closed bilateral multi issue setting using alternating offers negotiation protocol [2, 6]. The main challenge of ANAC 2014 was dealing with the nonlinear utility scenarios. In non-linear scenarios, the agents no longer have linear utility functions as was the case with the previous ANAC competitions in which the agents were negotiating in linear scenarios. Another challenge of this year was to deal with ultra large-size domains with outcome spaces as big as 50^{10} possible bids.

F. Zafari · F. Nassiri-Mofakham (✉)
Department of Information Technology Engineering, University of Isfahan,
Hezar Jerib Avenue, 81746-73441 Isfahan, Iran
e-mail: fnasirimofakham@yahoo.com; fnasiri@eng.ui.ac.ir
URL: http://eng.ui.ac.ir/~fnasiri/

F. Zafari
Faculty of Science, Engineering and Technology, Swinburne University of Technology,
Melbourne, VIC 3122, Australia
e-mail: f_z_uut@yahoo.com; f.zafari@eng.ui.ac.ir; fzafari@swin.edu.au

© Springer International Publishing Switzerland 2016
N. Fukuta et al. (eds.), *Recent Advances in Agent-based Complex
Automated Negotiation*, Studies in Computational Intelligence 638,
DOI 10.1007/978-3-319-30307-9_21

A negotiation strategy typically consists of four constituting parts: (1) Bidding Strategy, (2) Acceptance Strategy, (3) Opponent Model, and (4) Opponent Model Strategy. These parts work with each other within a BOA framework [1], to accomplish the negotiation task. Our Hybrid negotiation agent is based on the BOA framework, so we elaborate on the constituting parts of our agent separately, in the following sections.

Currently, three types of bidding strategies are used in the literature [4], namely: Time Dependent, Behavioral (or Imitative), and Random. Compared to the existing bidding strategies, BraveCat Agent uses a hybrid strategy in the sense that it gradually concedes over time, shows a form of randomness, and awards nice moves by the opponent. Our agent also uses a new opponent model, which is tailored to the negotiation setting of ANAC 2014. In ANAC 2014, the utility of a bid is no longer computed by a simple linear function. Instead, a complex nonlinear utility function is used for this purpose, so that its structure is unknown to the agents. So the conventional opponent models, such as frequency models and Bayesian models cannot be used to model the utility space or preference profile of the opponent. This limitation makes the modeling of the opponent's preferences even more challenging than before. So we design a new distance based opponent model and use it to estimate the utility of a candidate bid to be sent to the opponent in each round of the negotiation. To overcome the huge size of the negotiation domains, we employed iterative deepening search [8] which was also previously used by AgentK in ANAC 2010 [7].

The rest of the paper is organized as follows. In Sect. 2, we explain the acceptance strategy of our agent. In Sect. 3, we elaborate on our bidding strategy. This section describes how our agent identifies the final rounds of the negotiation process and how it exploits the obtained information to avoid break offs in a negotiation session. In Sect. 4, we explain the opponent model that our agent uses to increase the likelihood of the bids sent to be accepted by the opponent. And finally in Sect. 5, we conclude the paper.

2   Acceptance Strategy

The acceptance strategy of BraveCat Agent is comprised of two parts: (1) initial rounds, and (2) final rounds. As for the last round, the acceptance strategy is to accept the last bid unconditionally. The logic behind this is that the agent should avoid break offs (i.e., ending a negotiation session without an agreement) as much as possible. On the other hand, as most agents follow some kind of a concession strategy, the last bid would probably be as close to the best bid for agent, as possible. So BraveCat Agent recognizes the last round of the negotiation (please see Sect. 3.3.2) and accepts the last bid unconditionally. As for the initial rounds, the agent accepts if and only if the utility of that bid for the agent is greater than $U = 0.8$, the acceptable utility threshold. The value 0.8 was gained through experiments of BraveCat against a few agents from previous ANACs, adapted by ANAC organizers for a few nonlinear sample domains they had released for the setting of ANAC 2014 in Genius 5.1.

By increasing the value of U from 0.8 to 1, BraveCat might not be able to reach an agreement in the early times or even was not able to reach an agreement at all. When we decreased the value of U to 0.6, we could reach an agreement in the very early times of the negotiation, but we could gain just very poor utility value 0.6. So to satisfy this greediness for reaching an early but high utility agreement, BraveCat should strike a balance between increasing this acceptable utility threshold (and hence obtaining more utility values in the case of an agreement) and decreasing this acceptable utility threshold (and increasing the chance of making an agreement in the early negotiation rounds, instead). Using 0.8, BraveCat could make an agreement even at the early times of the negotiation while preserving a high utility.

3 Bidding Strategy

As mentioned in the first section, BraveCat Agent uses a hybrid bidding strategy, which is a mixture of a random strategy, a time dependent strategy, and a behavioral strategy. The proposed bidding strategy works as follows.

3.1 Time Dependent Behavior

First of all, a time dependent strategy produces a real number G_T, called target utility. This number is computed based on Eq. 1:

$$G_T = u_T(t) = 1 - t^{\frac{1}{\beta}} \tag{1}$$

in which, u_T is a time dependent utility function to compute the target utility, and t and β are time and the value that controls concession speed of the agent, respectively.

When β is equal to, less than, or greater than 1, then the agent behavior respectively will be Fixed Concession (i.e., linear), Conceder, and Boulware (Hardhead or Tough).

3.2 Random Behavior

Then, we randomize this time dependent strategy using a normal distribution[1] N(0, 1) and choose a random number z. Then, the new target value G_R will be obtained as Eq. 2:

---

[1] In normal distribution, when the random variable value resides in the interval $[-2.58, 2.58]$, the accumulative probability will equal to 99 %. That is, in a normal distribution experiment with 99 % probability, the random variable value will fall into the interval $[-2.58, 2.58]$.

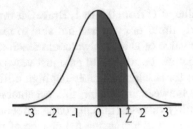

Fig. 1 Standard normal distribution probability curve

$$G_R = u_R(t) = u_T(t) - \frac{(1 - u_T(t)) \times |z| \times x}{2.58} \tag{2}$$

where, u_R is a randomized utility function and x determines the degree of its randomized/deterministic behavior. The z value in fact equals to the distance of the randomized target utility to the time dependent target utility (Fig. 1). Since we are interested in gradual transition from a random behavior to a deterministic behavior as the time passes in the negotiation, the probability of choosing a value close to the time dependent target utility should be higher in final times of the negotiation, compared to the initial times.

So we choose value x according to Eq. 3, in which γ controls the transition speed from a randomized to a deterministic behavior.

$$x(t) = 1 - t^{\frac{1}{\gamma}} \tag{3}$$

3.3 Imitative Behavior

The post tournament analysis of ANAC 2011 shows that Tit for Tat Agent was the only agent in the tournament that matched the behavior of the opponent [3, 6]. In other words, this agent plays tough against the hardheaded agents, while it plays nice against the agents with conceding bidding strategies. Although this strategy may be optimal against hardheaded strategies, it is not optimal against conceding strategies. Because when it plays against conceding agents, the optimal strategy may be to exploit their conceding behavior to reach a better score in the negotiation. So we need to strike a balance between cooperative and competitive behavior of the agent. For this purpose, we combine a time dependent tough strategy with a specific form of a behavioral strategy. The behavioral strategy we use is similar to Tit for Tat.

According to the Tit for Tat, the agent will initially cooperate, and then respond in kind to the opponent's previous action. That is, in a Tit for Tat strategy, the agent makes a nice move, as long as the opponent plays nice as well. As soon as the opponent defects, the agent retaliates with a defection. Then, as soon as the opponent makes a nice move after defection, the agent forgets about the past, and starts to play nice again. In other words, the agent's response is just a function of the opponent's last move. If the last move is a good one, the agent responds to that move with a nice move either. But if the last move is a defection, the agent responds to that move with a defection.

Our behavioral strategy is different in the way that it is only reciprocative not retaliative. Putting it differently, the agent tracks the moves made by the opponent, recognizes the nice moves and reciprocates them with a nice deviation from the target utility G_R computed in Eq. 2. The question is how is a nice move from the opponent defined? If the agent has complete information regarding the preference profile of the opponent, a nice move is a move which decreases the opponent's utility. Since the agent doesn't have complete information, a nice move is defined differently.

In every round of the negotiation, the agent keeps the best bid received from the opponent. Whenever a new bid is received, the agent checks if the utility of the new bid is greater than the best bid received so far. If so, a nice move is captured by the opponent. So the agent reciprocates this nice move, by temporarily changing its utility to the following utility in Eq. 4:

$$G_B = u_B(t) = u_R(t) - (u^{\text{Own}}(B[n(t)]) - u^{\text{Own}}(B[n(t) - 1])) \qquad (4)$$

where, G_B is the target value obtained using behavioral utility function u_B, n(t) denotes the number of best bids received from the opponent so far, B is a list containing the best bids received, and u^{Own} is the agent's utility function. Whenever a new bid is received, the agent compares u^{Own}(new bid) with u^{Own}(the last bid in this list). If u^{Own} (new bid) is better, the newly received bid is added to the list B (as the latest best bid), and a nice move by the opponent is identified. Then the agent reciprocates by conceding towards the opponent by using Eq. 4.

According to analysis of former ANACs, ability to accurately detecting the final rounds of the negotiation is a major factor in the success of a negotiating agent. Making strategic decisions at the final rounds of a negotiation is of paramount importance to the agent's success. Therefore, the BraveCat bidding strategy is comprised of two parts: (1) initial rounds, and (2) second-last round. The two constituting parts of the proposed bidding strategy are explained in the following sections.

3.3.1 Bidding in Initial Rounds

In the previous ANACs, it was possible to store all the bids in a list in the memory before starting the negotiation session, and then sort all the outcomes in the order of their utility values. Then, when deciding on a bid to propose to the opponent, the bidding strategy of the agent could easily find the bids with utility values equal to or greater than the target utility value of the agent (computed using Eq. 4). The agent then could choose one of those bids to propose to the opponent. However, in ultra large domains this would not be possible to do due to memory limitations imposed by the huge amount of memory needed. To overcome the challenge of making bids

```
temporary variables:
θ denotes target utility value,
ε denotes a very small amount,
n denotes mximum number of iterations,
  While true
              i = 0 ;
          While i ≤ n
                  Generate a Random a Bid;
                  Calculate Utility Value of the Bid;
                  If Utility of the Bid ≥ θ
                        Output the Bid;
                        Break;
                  EndIf
          EndWhile
              θ = θ - ε;
              i = i + 1;
  EndWhile
```

Fig. 2 BraveCat iterative deepening search for an acceptable bid for the agent

in ultra large domains, we adopt an iterative deepening search previously used by
AgentK in ANAC 2010 [7].

For the initial rounds, the agent chooses the best bid among all acceptable bids
for the agent. A bid is acceptable for the agent, if and only if its utility is greater than
or equal to the target utility value of the agent at that time (calculated according to
Eq. 4). The best acceptable bid is the bid having the greatest estimated utility value,
which is obtained from the opponent model. Figure 2 shows the algorithm BraveCat
employs for finding an acceptable bid in huge negotiation domains.

3.3.2 Recognizing the Final Rounds

As shown in Fig. 3, the agent receives a bid, and later at the time t should decide on
the message it is going to send. This round is the last round of the negotiation, if and
only if the negotiation ends before the agent receives the next bid. To estimate the
next bid receiving time, we use the time elapsed between the last bid received and
the bid received before it (denoted by Y in Fig. 3). Then, we expect the next bid to
be received in the time t_e computed as Eq. 5:

$$t_e = t_l + 0.004 \times n \times Y \tag{5}$$

where, t_l and n are the time the last bid was received and the number of negotiation
rounds passed so far, respectively. Through experimental sessions we ran to compare
BraveCat in ANAC 2014 sample nonlinear domains against sample agents, we found
that the value 0.004 resulted in less number of break-offs.

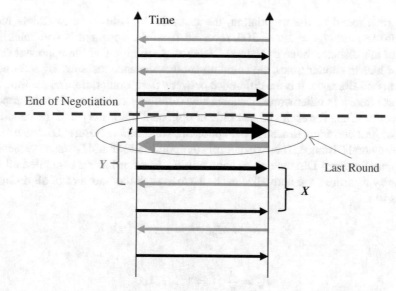

Fig. 3 Conceptual plot showing the temporal position of the last round in a bilateral negotiation session

So if the next bid's receiving time exceeds the negotiation deadline, the last bid received so far (the thick left arrow circled in Fig. 3), would probably be the last bid the agent would receive through the negotiation session. Therefore, the last message sent in this negotiation will be an Accept Message. When sending a bid, if the agent finds out that the current bid would probably be one of the last bids sent, it will choose to send the best bid received from the opponent so far. This is reasonable since the bid sent by the opponent at early times, would also probably be acceptable at later times, and on the other hand, it is the best utility the agent could obtain in the final rounds of the negotiation.

4 Opponent Model

The proposed agent was designed to participate in nonlinear scenarios where we do not have any information regarding the exact structure of the utility function which is used to compute the utility of a bid. Therefore, conventional methods to model the preferences of the opponent such as Frequency models [9] or Bayesian models [5] are useless in such scenarios. So we need to design a specific model for nonlinear scenarios.

For this purpose, we assume that two bids having near Euclidean Distance values would probably have similar utilities. So we designed and implemented a new Distance based model for nonlinear scenarios and incorporated this model into the BraveCat Agent. This method works as follows.

In each round of the negotiation, the distance N_b^i between a candidate bid b and 100 last bids $i, i = 1, \ldots, 100$, received from the opponent is computed. We refer to this distance value as Natural Distance. Assuming that the opponent sends similar bids in similar times, we compute another distance measure M_b^i referred to as Temporal Distance. It is the difference between the current time, and the time that bid was offered. In other words, it shows how different two bids are from a temporal perspective. Then for each 100 last bids in opponent bid history and the candidate bid the agent considers to send to the opponent, we compute Natural Distance along with Temporal Distance. Then we multiply temporal and Natural Distance values and we obtain the Final Distance F_b^i as shown in Eq. 6. After having computed all 100 similarity measures, we normalize each distance such that the sum of all distances equals to 1.

$$F_b^i = N_b^i \times M_b^i, \quad i = 1, \ldots, 100 \tag{6}$$

$$P_b^i = \frac{1}{F_b^i}, i = 1, \ldots, 100 \tag{7}$$

where P_b^i corresponds to the probability to which the utility of the candidate bid b is equivalent to the utility of the opponent bid.

Then, we assume that the opponent follows a concession based strategy. This assumption is pretty common in the literature [4, 5]. So we assume that the opponent sends bids with utility $(1 - 0.3t)$, $0 \le t \le 1$, during the negotiation session. Then, by multiplying each estimated utility value of the opponent for each bid to its corresponding probability and summing the obtained values for all the 100 bids, the estimated utility u^{OP} of the candidate bid is obtained as shown in Eq. 8:

$$u^{OP}(b) = \sum_{i=1}^{100} (1 - 0.3t(i)) \times P_b^i, \quad 0 \le t \le 1 \tag{8}$$

where, $t(i)$ is the time at which the last i-th bid is received from the opponent.

5   Conclusion

In this paper, we briefly proposed BraveCat Agent, which was mainly designed to take part in the international negotiating agents competition (ANAC) 2014 and qualified to enter the final rounds of the competition. Unlike previous ANACs in which the agents were negotiating on linear scenarios, in ANAC 2014, nonlinear utility functions are used to calculate the utility of a bid. Another challenge of this year was to deal with ultra large-size domains with outcome spaces as big as 50^{10} outcomes. Based on BOA framework [1], BraveCat was comprised of bidding strategy, opponent model, and acceptance strategy components. BraveCat hybrid bidding strategy combined a time dependent bidding strategy, a random strategy, and an imitative strategy. A new opponent model based on Euclidean Distance was also used to model the preferences of the opponent in nonlinear scenarios. To prevent the negotiation break offs, we predicted the last round of the negotiation by using the current number of rounds elapsed and the time length of the last round. BraveCat could overcame the memory limitations regarding bidding in ultra large-size domains by using iterative deepening search for the acceptable bids for the agent.

References

1. Baarslag, T., Hindriks, K., Hendrikx, M., Dirkzwager, A., Jonker, C.: Decoupling negotiating agents to explore the space of negotiation strategies. In: Novel Insights in Agent-based Complex Automated Negotiation, pp. 61–83. Springer (2014)
2. Baarslag, T., Hindriks, K., Jonker, C., Kraus, S., Lin, R.: The first automated negotiating agents competition (anac 2010). In: New Trends in Agent-based Complex Automated Negotiations, pp. 113–135. Springer (2012)
3. Baarslag, T., Hindriks, K., Jonker, C.: A tit for tat negotiation strategy for real-time bilateral negotiations. In: Ito, T., Zhang, M., Robu, V., Matsuo, T. (eds.) Complex Automated Negotiations: Theories. Models, and Software Competitions, volume 435 of Studies in Computational Intelligence, pp. 229–233. Springer, Heidelberg (2013)
4. Faratin, P., Sierra, C., Jennings, N.R.: Negotiation decision functions for autonomous agents. Robot. Auton. Syst. 24(3), 159–182 (1998)
5. Hindriks, K., Tykhonov, D.: Opponent modelling in automated multi-issue negotiation using bayesian learning. In: Proceedings of the 7th international joint conference on Autonomous agents and multiagent systems, pp. 331–338. International Foundation for Autonomous Agents and Multiagent Systems, Estoril, Portugal (2008)
6. Katsuhide, F., Ito, T., Baarslag, T., Hindriks, K., Jonker, C., Kraus, S., Lin, R.: The second automated negotiating agents competition (anac2011). In: Ito, T., Zhang, M., Robu, V., Matsuo, T. (eds.) Complex Automated Negotiations: Theories. Models, and Software Competitions, volume 435 of Studies in Computational Intelligence, pp. 183–197. Springer, Heidelberg (2013)
7. Kawaguchi, S., Fujita, K., Ito, T.: Agentk: compromising strategy based on estimated maximum utility for automated negotiating agents. In: Ito, Takayuki, Zhang, Minjie, Robu, Valentin, Fatima, Shaheen, Matsuo, Tokuro (eds.) New Trends in Agent-Based Complex Automated Negotiations. Studies in Computational Intelligence, vol. 383, pp. 137–144. Springer, Heidelberg (2012)
8. Russell, S., Norvig, P.: Artificial Intelligence: A Modern Approach (2010)
9. van Krimpen, T., Looije, D., Hajizadeh, S.: Hardheaded. In: Ito, T., Zhang, M., Robu, V., Matsuo, T. (eds.) Complex Automated Negotiations: Theories. Models, and Software Competitions, volume 435 of Studies in Computational Intelligence, pp. 223–227. Springer, Heidelberg (2013)

Printed in the United States
By Bookmasters